QA 279.4 S76 2012 C.2

Theory of Conditional Games

Game theory explains how to make good choices when different decision makers have conflicting interests. The classical approach assumes that decision makers are committed to making the best choices for themselves regardless of the effect on others, but such an approach is less appropriate when cooperation, compromise, and negotiation are important. This book describes conditional games, an extension of game theory that accommodates multiple stakeholder decision-making scenarios where cooperation and negotiation are significant issues and where notions of concordant group behavior are important. Using classical binary preference relations as a point of departure, the book extends the concept of a preference ordering that permits stakeholders to modulate their preferences as functions of the preferences of others. As these conditional preferences propagate through a group of decision makers, they create social bonds that lead to notions of group concordance.

This book is intended for all students and researchers of decision theory and game theory, including students in artificial intelligence (especially multiagent systems and distributed control), economics, management science, psychology, analytic philosophy, and applied mathematics.

WYNN C. STIRLING is a professor of Electrical and Computer Engineering and Dean of Graduate Studies at Brigham Young University. His current research interests include multiagent decision theory, estimation theory, information theory, and stochastic processes. He is the author or co-author of more than seventy publications, including the graduate text *Mathematical Methods and Algorithms for Signal Processing* (with Todd Moon) and the monograph *Satisficing Games and Decision Making*.

THEORY OF CONDITIONAL GAMES

WYNN C. STIRLING

Brigham Young University

CAMBRIDGE
UNIVERSITY PRESS

CAMBRIDGE UNIVERSITY PRESS
Cambridge, New York, Melbourne, Madrid, Cape Town,
Singapore, São Paulo, Delhi, Tokyo, Mexico City

Cambridge University Press
32 Avenue of the Americas, New York, NY 10013-2473, USA

www.cambridge.org
Information on this title: www.cambridge.org/9781107011748

First published 2012

Printed in the United States of America

A catalog record for this publication is available from the British Library.

Library of Congress Cataloging in Publication data
Stirling, Wynn C.
Theory of conditional games / Wynn C. Stirling.
p.   cm.
Includes bibliographical references and indexes.
ISBN 978-1-107-01174-8
1. Decision making.   2. Games of strategy (Mathematics)   I. Title.
QA279.4.S76 2012
519.3–dc23   2011041577

ISBN 978-1-107-01174-8 Hardback

To Tony, Dave, and Kate

Natural science is an expansion of observing; technology, of contriving; mathematics, of understanding.
      — Michael Polanyi
        Personal Knowledge (University of Chicago Press, 1958)

Contents

Preface

Hypothesen sind Netze, nur der wird fangen, der auswirft.
Theories are nets: only he who casts will catch.
— *Novalis (Friedrich von Hardenberg)*
Dialogen und Monolog, 1798

John Dewey observed that "In scientific inquiry, the criterion of what is taken to be settled, or to be knowledge, is being *so* settled that it is available as a resource in further inquiry; not being settled in such a way as not to be subject to revision in further inquiry" [emphasis in original, Dewey, 1938, pp. 8–9]. Game theory has been successfully applied to the subject matter of general economic theory, particularly for competitive and market-driven scenarios where individual rationality dominates. It is firmly established; it is settled. In fact, it is *so* settled that it is available as a resource for further inquiry into the issue of multistakeholder decision making for scenarios where social relationships extend beyond self-interest.

The net cast by classical game theory, however, is designed to capture the essential characteristics of multistakeholder decision scenarios where all participants possess categorical (unconditional) preference orderings and are committed to achieving the best individual outcomes for themselves. This book revises game theory to cast a wider net designed to capture, in addition, decision-making scenarios where cooperation, compromise, negotiation, and altruism are significant issues, and where notions of concordant group behavior are important. Using classical binary preference relations as a point of departure, the book extends the concept of a preference ordering to permit stakeholders to modulate their preferences as functions of the preferences of others. As these conditional preferences propagate through a group of autonomous decision makers, they create social bonds that lead naturally to notions of group concordance. This concept is formalized by defining a conditional game as a set of players, their feasible actions, and their conditional preference orderings.

Conditional game theory permits the development of (a) solution concepts that simultaneously account for both individual preferences and group concordance; (b) negotiation protocols that permit each stakeholder to achieve its security level while maximizing the concordance of the group; (c) a theory to characterize the intrinsic ability of a group to coordinate; (d) a mechanism to combine the deterministic and stochastic components of decision making into a single unified utility structure; and (e) rigorously defined concepts of multiple-agent satisficing that relax the brittle demands of optimization and provide decision makers with the flexibility to achieve solutions that accommodate multiple points of view.

Motivation

Historically, game theory has mainly been the purview of the social sciences (chiefly economics and political science). Recently, however, engineering and computer science have employed game theory as a useful framework within which to model multiagent decision making and control. Thus, the book is written to both the social science and the engineering/computer science audiences. To appeal to multiple disciplines, it focuses on the foundational assumptions that underlie the formulation of decision-making methodologies and does not delve deeply into practical applications or case studies.

Although they share similar mathematics, social science and engineering/computer science differ significantly in the way game theory is used. On the one hand, social science uses game theory as an *analysis* tool with which to explain, predict, justify, and recommend human action. But the solution does not dictate action – it is not causal. At best it is a reasonable approximation under controlled circumstances. When observed behavior deviates from the behavior predicted by the game, it is necessary to invoke ex post psychological or sociological arguments that, while not part of the mathematical model, are necessary to explain the behavior.

On the other hand, engineers and computer scientists use game theory for *synthesis*: to design and construct artificial decision-making agents. When used this way, the solution does indeed dictate behavior – it is causal. Whereas a model used for analysis can be a convenient fiction, a model used for synthesis creates its own reality. A multiagent system cannot rely on ex post psychological or sociological interpretations to explain inappropriate behavior. In particular, if notions of social behavior are relevant, then they must be explicitly encoded into the model.

The intended readership

This book is intended for all students of decision theory and game theory. Researchers and students in distributed artificial intelligence, distributed control, and multiagent systems will find the material covered in this book to be quite different from the treatments that are directed to computer science and engineering audiences, and thus may supplement standard game theoretic–based treatments as applied to those disciplines. Also, researchers and students in the social sciences will benefit from exposure to a perspective of game theory that is not heavily emphasized in the conventional social science literature, namely, extending beyond considerations of individually rational behavior. In addition, since this book touches on such philosophical issues as rationality and sociality, it should be of interest to philosophers and other students of rational choice theory.

Structure of the book

Chapter 1 moves beyond classical notions of rationality and argues that social influence relationships that exist among stakeholders must be accounted for explicitly, and that concepts of individual and group rationality must be reconciled if game theory is to be successfully applied to engineering and artificial intelligence decision problems.

Chapter 2 extends game theory by replacing ex ante categorical preference orderings, which unconditionally characterize each stakeholder's interests before the game is played, with preference orderings that permit stakeholders to condition their preferences on the preferences of others, thereby dynamically modulating their preferences as the game is played. These conditional preferences propagate throughout the group to create a social bond among its members, thereby making it possible to define preference orderings for both the individuals and the group. If the preferences of all stakeholders are categorical, however, then no notion of group preference exists, and the situation reverts to a conventional game. Thus, conditional game theory is a true generalization of the classical theory.

The extension to conditional games makes it possible to introduce theoretical developments that are not permitted within the classical structure. **Chapter 3** provides an expanded solution concept that simultaneously accounts for both group and individual interests. This new concept provides a framework for the negotiation of a compromise that allows each agent to reconcile its own interests with the interests of others and the group.

Chapter 4 introduces a formal notion of coordinatability. Drawing on concepts from Shannon information theory, it presents a notion of coordination capacity that provides a quantified measure of the intrinsic ability of a group to function in a coherent, harmonious way, independently of the notions of rational behavior that are adopted by the players. The coordination capacity essentially provides an ecological measure of how well an organization is able to function in its environment.

Chapter 5 offers a unified treatment of the preferential and probabilistic aspects of a multistakeholder decision problem in the presence of uncertainty. The result is the definition of a combined utility-probability network that possesses all of the mathematical properties of a standard Bayesian network. This structure enables the symmetric representation of the deterministic and random components of a decision problem, which facilitates the calculation of expected utility.

Chapter 6 generalizes the notion of satisficing game theory, which was first presented in my earlier book, *Satisficing Games and Decision Making* (Cambridge University Press, 2003). This theory presents an alternative to the classical approach of seeking the best and only the best solution to a multistakeholder decision problem by defining a set of solutions that satisfy a rigorous notion of what it means to be good enough. This approach provides increased flexibility in choosing a solution that is acceptable to both the individuals and the group.

Chapter 7 provides a number of examples that illustrate the use of conditional game theory, and **Chapter 8** is an invitation to explore the theory of conditional games more deeply, with the hope that other researchers will contribute to its further theoretical development and to useful applications.

The usage of personal pronouns

Depending on the context, a stakeholder may be a person, an artificial agent, an attribute, or a random phenomenon. To accommodate this diversity, I have adopted the convention of using "it" for third-person singular pronouns unless the gender of the antecedent is obvious from the context.

Acknowledgments

Developing the ideas expressed in this book has been a joy. Although it is a relief to bring this project to conclusion, I cannot help but reflect on the sentiment expressed by Gauss: "It is not knowledge, but the act of learning, not possession but the act of getting there, which grants the greatest enjoyment." The seed ideas for this book arise from what may seem to be an unlikely source for an engineer: the writings of the distinguished epistemologist Isaac Levi, who insists that attention should be focused on the improvement of knowledge, not merely its justification. I have transmigrated his theory of epistemic utility from the epistemological domain to the praxeological domain, and have extended it to the multiagent (multivariate) case. His emphasis on formulating a mathematical framework for dynamically synthesizing knowledge resonates with the engineering concept of designing and synthesizing mathematical decision-making devices.

Many colleagues have contributed to the ideas contained in this book. Of particular note are three of my former graduate students, Darryl Morrell, Michael Goodrich, and Matthew Nokleby, whose creativity and enthusiasm have sustained me many times, and whose ideas are integral to this theory. I have also greatly benefited from my collaborations with Richard Frost, Harold Miller, Todd Moon, Dennis Packard, and Teppo Felin. The research environment and support offered by Brigham Young University and, particularly, by my colleagues in the Electrical and Computer Engineering Department, who have graciously tolerated my unorthodox research agenda, are greatly appreciated. A special thanks goes to Laura Rawlins of BYU's Faculty Editing Service, whose careful reading of the manuscript provided many much needed clarifications, and to Lauren Cowles and David Jou of Cambridge University Press for their support and patience.

My greatest debt, however, is to my wife Patti, whose love has sustained and nourished me for more than four decades. Her special insights and wisdom contribute an intangible, but nevertheless essential, element to this enterprise.

WYNN C. STIRLING
Provo
May 2011

1

Sociality

One can argue very persuasively that the weakest link in any chain of
argument should not come at the beginning.
— *Howard Raiffa*
Decision Analysis (Addison-Wesley, 1968)

Decision making is perhaps the most fundamental intellectual enterprise.
Indeed, the word *intelligent* comes from the Latin roots *inter* (between) +
legĕre (to choose). The study of motives and methods regarding how decisions
might and should be made has long captured the interest of philosophers and
social scientists and, more recently, of engineers and computer scientists. An
important objective of such studies is to establish a framework within which to
define rational behavior and solution concepts that result in appropriate choices.
The development of formal theories of decision making, however, has proven
to be a challenging and complex task. The reason is simple: Every nontrivial
decision problem involves multiple stakeholders. A stakeholder is any entity
that has an interest in the consequences of a decision, whether or not it has
direct control over the decision. Unless the interests of all stakeholders coincide perfectly (a rare occurrence), conflicts will exist. The central challenge
of any theory of decision making, therefore, is how to make choices in the
presence of conflicting stakeholder interests.

   The way a group of stakeholders deals with conflict is a function of its sociality: Conflict can result in either competition or cooperation. At one extreme
(speaking anthropomorphically), each member of a group views others as adver-
saries and treats them antagonistically. At the other extreme, it views others as
partners and treats them synergistically. On the one hand, others are viewed as
opponents that can only constrain achieving one's selfish desires; on the other
hand, others are viewed as teammates with whom to pursue common objectives that cannot be achieved effectively alone. Members of a group typically

1

fall between these two extremes, displaying mixed motives, as they balance opportunities to benefit themselves regardless of the expense to others with opportunities to benefit others at possibly their own expense.

Any systematic account of rational decision making requires that a mathematical model be specified that defines the stakeholders, their possible alternatives, their preferences over the alternatives, and their concepts of rational behavior.[1] Game theory, as developed by von Neumann and Morgenstern (1944) is perhaps the best known such mathematical model. In this treatment, we restrict attention to finite strategic (normal form) single-stage noncooperative games.[2] Formally, a game consists of two or more autonomous stakeholders, or players, each of whom has a finite set of pure strategies (deterministic actions in a single-stage context) from which it may choose one action to instantiate. An action profile is an array of actions, one for each player. These profiles constitute the outcomes, or consequences, of the game. Each player also possesses a preference ordering over the outcomes. Typically, these preference orderings are expressed in terms of numerical valuations, called payoffs or utilities, that define the benefits (either ordinally or cardinally) to the players.

Representing the decision problem as a mathematical game permits the players to strip the decision problem from its context and to examine it dispassionately from the point of view of possible actions and outcomes. Most developers of such models have faithfully adhered to Occam's razor and have resisted the introduction of complicating factors that are not deemed essential. When defining a game, however, it is imperative also to consider what some have termed Einstein's razor: "It can scarcely be denied that the supreme goal of all theory is to make the irreducible basic elements as simple and as few as possible without having to surrender the adequate representation of a single datum of experience" (Einstein, 1934, p. 165).[3] Seemingly, everyone can agree with the first part of this dictum to keep things simple. But it is the latter injunction, not to surrender adequate representation, that is perhaps more difficult to accommodate.

Once the stakeholders and actions are specified, two fundamental elements remain to be defined when formulating a multistakeholder decision problem: (a) the structure of the preference orderings over the outcomes and (b) the notions of rationality that are used to formulate solution concepts.

---

[1] Henri Poincaré observed that "mathematics is a language by which no indistinct, obscure, or indefinite things can be expressed" (cited in Mérö (1998, p. 230).)

[2] For many scenarios, a multistage game can be described by a series of single-stage games, particularly when the key issue of concern is coordination.

[3] This quote has sometimes been reworded into such variants as "Everything should be made as simple as possible, but not simpler."

By far the most prevalent assumption employed by decision theory when considering preference orderings is also the most simple: A preference ordering over outcomes is well defined for each individual stakeholder. Arrow (1951) put it succinctly: "It is assumed that each individual in the community has a definite ordering of all conceivable social states, in terms of their desirability to him. . . . It is simply assumed that the individual orders all social states by whatever standards he deems relevant" (p. 17). According to this view, each stakeholder's preference ordering is completely and immutably defined by the payoffs before the stakeholders engage in the act of making choices. Such a preference ordering is *categorical*: It unconditionally defines the stakeholder's valuation system under all circumstances, regardless (at least ostensibly) of the valuations of other stakeholders.

The second fundamental element of a decision model is the concept of rational behavior that governs the way stakeholders use the information at their disposal. The simplest possible rationality model, which is also the most widely used, is that each member of a group will restrict interest to its own benefit and will act in a way that achieves its best possible outcome. This is the doctrine of *individual rationality*. As observed by Tversky and Kahenman (1986),

> The assumption of [individual] rationality has a favored position in economics. It is accorded all of the methodological privileges of a self-evident truth, a reasonable idealization, a tautology, and a null hypothesis. Each of these interpretations either puts the hypothesis of rational action beyond question or places the burden of proof squarely on any alternative analysis of belief and choice. The advantage of the rational model is compounded because no other theory of judgment and decision can ever match it in scope, power, and simplicity. (p. 89)

The mathematical structure of categorical preference orderings and the logical structure of individual rationality are ideally matched to each other. Harsanyi (1977) articulated this point as follows: "Because all values and objectives in which the players are interested have been incorporated into their payoff functions, our formal analysis of any given game must be based on the assumption that each player has only one interest in the game – to maximize his own payoff" (p. 13). Given categorical preference orderings, the only compatible notion of rationality is self-interest, since the preferences are restricted to, and only to, individual welfare. Conversely, given individual rationality, any structure other than categorical preferences would extend interest beyond the self. With these structures, it is possible to formulate precise definitions of rational behavior for the members of a group.

What is lacking with this structure, however, is a concept of preference for the group and, hence, a concept of rational behavior for the group. One reason for the lack of focus on group preferences is that since the group, as an entity, does not possess the capability to enforce a decision even if it had a preference, the relevance of a group preference ordering is problematic. Furthermore, as Arrow's (1951) impossibility theorem establishes, it is not generally possible to define a preference ordering for a group by simply aggregating individual preferences of its members without violating a set of arguably reasonable and desirable properties. Thus, although it is the joint decision of the members of the group that defines the outcome and, hence, the benefits to the individuals within the group, the group itself does not possess a preference ordering. As argued by Luce and Raiffa (1957), "the notion of group rationality is neither a postulate of the model nor does it appear to follow as a logical consequence of individual rationality" (p. 193). Consequently, classical decision theory has proceeded by making assumptions about individual preferences only and then using those preferences to deduce information about the choices (but not the values) of a group.

The classical game-theoretic approach of focusing exclusively on individual preferences may be justified when the members of a group fit the model; that is, when they really are able to define their preferences categorically and are motivated by, and only by, self-interest. But that is an extreme situation; it is an abstraction that must be justified in application, not merely taken as a tenet of a classical doctrine to be applied uncritically. Arrow (1986) clearly delimits the context in which this model applies:

> Rationality in application is not merely a property of the individual. Its useful and powerful implications derive from the conjunction of individual rationality and other basic concepts of neoclassical theory – equilibrium, competition, and completeness of markets....When these assumptions fail, the very concept of rationality becomes threatened, because perceptions of others and, in particular, their rationality become part of one's own rationality. (p. 203)

Despite Arrow's (1986) concern, the classical model of categorical preferences and individual rationality is routinely applied in contexts where, in addition to competition, opportunities for cooperation, compromise, and unselfishness are present. The application of the classical model in such contexts can lead to paradoxes and dilemmas that, although interesting and even charming, may be evidence that the model does not adequately account for the social relationships that exist among the members of the group.

Maslow observed that if the only tool you have is a hammer, you tend to see every problem as a nail. The classical game theory model may be a good hammer to drive the nail of competitive and market-driven decision making, but it may not be the best tool to model scenarios where more sophisticated social relationships exist. Although this tool has been effective in economic contexts, as the applications of multistakeholder decision making expand into other domains, its limitations become more pronounced. In his musing about the history of decision making, Shubik (2001) commented on the limitations of the classical approach to multistakeholder decision making:

> Economic man, operations research man and the game theory player were all gross simplifications. They were invented for conceptual simplicity and computational convenience in models loaded with implicit or explicit assumptions of symmetry, continuity, and fungibility in order to allow us (especially in a pre-computer world) to utilize the methods of calculus and analysis. Reality was placed on a bed of Procrustes to enable us to utilize the mathematical techniques available. (p. 4)

There are essentially two ways to address the limitations of the classical model: One may retrofit the old bed of categorical preferences and individual rationality, or one may create a new framework that is designed to deal explicitly with social relationships that are more complex than competition, including cooperation, compromise, and negotiation. Pursuing the former approach simply means continuing to force a group with a complex social structure into the classical preference structure/rationality bed; with the latter approach, the goal is to make the bed a better fit for its occupant.

The central theme of this book is to present a new model that (a) permits individuals to modulate their preference orderings to accommodate the interests of others as well as themselves and (b) employs notions of rationality that simultaneously apply to both individuals and the group. This new model sits at the intersection of two diverse disciplines: the social sciences, including economics, psychology, sociology, and political science, and the engineering and computer science disciplines (including distributed artificial intelligence, intelligent control theory, and multiagent systems theory). Although they have much in common mathematically, these two disciplines have different application scenarios. In the social sciences, models are used for analysis – that is, to explain, predict, justify, and recommend choices for human societies – but for engineering and computer science, models are used for synthesis – that is, to design and construct artificially intelligent decision-making societies such as multiagent systems and distributed control systems that must function autonomously. Whereas models used in the social sciences to characterize human behavior are noncausal and serve only as approximations to reality, models used in the engineering and computer science contexts are causal – they dictate behavior and create reality. It is critical, therefore, that such models must be capable of explicitly accounting for complex social relationships when they exist (the analysis context) or when they are desired (the synthesis context).

The intent of this presentation is to supplement, rather than supplant, existing theory and methodology. However, since this theory challenges some of the closely held assumptions that have served for decades as the foundational tenets of multistakeholder decision theory, it is important to review these assumptions, discuss their limitations, and lay the framework for going beyond them.

1.1 Classical theory

The doctrine of individual rationality has had enormous influence in the formulation of decision theories. One of the main justifications for this doctrine is its apparent consistency with the evolutionary theory of natural selection. The basic idea is that selfish characteristics evolve because they make the individual more fit for survival. Thus, natural selection promotes egoism. Furthermore, the argument goes, natural selection inhibits altruism, since helping others to survive would likely diminish the individual's own chances for surviving.

Nevertheless, the stubborn fact of ostensibly altruistic behavior in human society is readily observed. Examples abound of people sacrificing their own interests to benefit others. But sacrificing one's own interest to benefit another can also be viewed as fundamentally egoistical (people act in that way to feel good about themselves). At the end of the day, however, arguments that self-interest is the primary motive for human behavior are inconclusive. Sober and Wilson (1998) sum it up this way:

> Psychological egoism is hard to disprove, but it also is hard to prove. Even if a purely selfish explanation can be imagined for every act of helping, this doesn't mean that egoism is correct. After all, human behavior also is consistent with the contrary hypothesis – that some of our ultimate goals are altruistic. Psychologists have been working on this problem for decades and philosophers for centuries. The result, we believe, is an impasse – the problem of psychological egoism and altruism remains unsolved. (pp. 2–3)

If the problem is unresolved, it would be prudent to refrain from relying exclusively on a rationality model that is based solely on self-interest, either when analyzing human behavior or, especially, when designing artificially intelligent entities that are intended to work harmoniously, and perhaps altruistically, with each other and with humans.

As a motive for action, self-interest is perhaps the most common justification that is invoked by decision theory for behavior in multistakeholder contexts. Without doubt, it is the simplest motive imaginable, since it takes into consideration only the benefit to the individual. Nevertheless, the very concept of self-interest, as a well-defined motive, has come under criticism. The key argument is that the concept is so simple that it is essentially vacuous. As argued by Holmes (1990), "The decision to group together sharply dissimilar motives under the single category of 'calculating self-interest' is said to involve an undesirable *loss of information* about rudimentary psychological and behavioral processes.

This is the essence of Macaulay's mocking remark that to discover self-interest behind an action is to say, with tautological banality, that 'a man had rather do what he had rather do' (Macaulay, 1978) [emphasis in original][4] (p. 269).

On the other hand, advocates of the self-interest concept argue that, while perhaps oversimplified, it nevertheless enables the construction of mathematical models to characterize behavior, and serves a prescriptive, even if not necessarily descriptive, role in the formation of a quantitatively based theory of value and behavior. As noted by Hogarth and Reder (1986),

> The role of [individual] rationality is to provide a principle (or "rationale") to mediate the relations between changes in one or more resource constraints and changes in the quantities of the relevant phenomena. This takes the form of the maintained hypothesis that *each of the individual decision makers behaves as if he or she were solving a constrained maximization problem* [emphasis added]. (p. 3)

The issue is further clouded by the existence of two flavors of self-interest: *narrow self-interest* and *enlightened self-interest*. With the former, the individual defines its preferences in accordance with its own welfare, and only its own welfare, regardless of the effect on others. With the latter, the individual defines its preferences in a way that improving the welfare of others ultimately improves its own welfare. This distinction may be important when considering the process of how one defines one's preference ordering, but once encoded into categorical preference orderings, self-interest is self-interest, regardless of the modifier, and the decision is made dispassionately according to whatever solution concept is applied.

There are no restrictions on the criteria that an individual uses to define its own self-interest. Although there is no explicit mechanism for an individual to consider the interests of others, the individual can always simulate the interests of others by substituting those interests in lieu of its own. As Sen (1990) observed:

> It is possible to define a person's interests in such a way that no matter what he does he can be seen to be furthering his own interests in every isolated act of choice.... No matter whether you are a single-minded egoist or a raving altruist or a class-conscious militant, you will appear to be maximizing your own utility in this enchanted world of definitions. (p. 19)

Once self-interest is redefined, a new game is created, and the issue devolves to considerations of which game is to be played. (In fact, one could randomly

---

[4] There is an interesting parallel between self-interest and survival of the fittest. "The semantic emptiness of the doctrine [survival of the fittest] was long ago exposed by asking the simple question, 'Fittest for what?' The only possible answer is 'fittest to survive,' which closes the circle and thereby reduces the statement to complete nonsense by making it read: 'Those survive who survive'" (Krutch, 1953, p. 85).

choose between the two games in a desperate attempt to reconcile one's conflicted interests, but such an approach would only obfuscate the issue further
and delay a more comprehensive approach.) The only reasonable assumption
from the point of view of classical game theory is that each player's payoffs
defines its true preferences.

If an individual has concerns for the welfare of others as well as itself,
restricting it to a categorical preference ordering severely limits its ability to
characterize an enlarged sphere of interest. This restriction may be appropriate
when self-interest is the operative attribute, but it does not easily account for
such attributes as concern for others, willingness to cooperate, and moral principles. When observed behavior deviates from the behavior predicted by the
model, it is necessary to invoke psychological and sociological attributes that,
while not part of the mathematical model, are necessary to explain the deviations. They merely overlay the basic mathematical structure of a game and avoid
or postpone a more profound solution, namely, the introduction of a model structure that explicitly accounts for complex social relationships and accommodates
notions of rational behavior that extend beyond narrow self-interest.

1.2 Sociality models

Homans (1961) offers three criteria for behavior to qualify as *social*. First, an
individual's actions must elicit some form of reward or punishment as a result
of behavior by another individual. Second, behavior toward another individual
must result in reward or punishment from that individual, not just a third party.
Third, the behavior must be actual behavior, not just a norm of behavior.

1.2.1 Minimal sociality

Under the paradigm of individual rationality, the stakeholders are indifferent, at
least ostensibly, to the welfare of others; they are manifestly egocentric. Even
so, it can be the case that some notion of social behavior can emerge as a result
of strategic reasoning, such as the attainment of a Nash equilibrium, a strategy
profile such that if any player unilaterally deviates, then its payoff is reduced.
A Nash equilibrium is a solution to a constrained optimization problem: Each
player does the best for itself, assuming that all other players are similarly
motivated. Such behavior, however, benefits only the individuals; benefit to
the group is undefined. Thus, although the behavior is a minimal expression of
sociality, a concept of group-level welfare is nonexistent.

Table 1.1. *The payoff matrix in ordinal form for the Prisoner's Dilemma game*

	X_2	
X_1	C	D
C	(3, 3)	(1, 4)
D	(4, 1)	(2, 2)

Key: 4 = best; 3 = next-best; 2 = next-worst; 1 = worst

Example 1.1 Perhaps the most well known of all games is the Prisoner's Dilemma (PD). The players, denoted X_1 and X_2, each have two possible actions: cooperate (C) or defect (D). Conventionally, the game is defined by an ordinal payoff matrix of the form displayed in Table 1.1. This game serves as a model for situations where mutual cooperation is better than mutual noncooperation for both players, but unilaterally attempting to cooperate leaves one player vulnerable to exploitation by the other. If one player chooses to cooperate and the other defects, the one who attempts to cooperate receives the worst payoff, while the other receives the best payoff. The dilemma arises because mutual defection (D, D) is the unique Nash equilibrium, and, according to the doctrine of individual rationality, the players should adopt this pessimistic next-worst solution rather than the more optimistic next-best (Pareto optimal) mutual cooperation solution (C, C). □

Regardless of the choices that are made, there is no clearly defined notion of group preference for the Prisoner's Dilemma; that is, the preference of the group is viewed as a whole and not individually. Of course, an external party is free to ascribe a notion of group preference, such as arguing that the group as a whole is better off if both choose to cooperate, but such an exogenously ascribed notion would be arbitrary.

An argument can also be made that mutual cooperation can emerge as a result of repeatedly playing the PD game, and that such behavior can be viewed as a group-level notion of preference. If the game is played an indefinite or a random number of times, the incentive to defect can be overcome by the threat of punishment, and mutual cooperation can emerge as an equilibrium, as attested by the many experiments and contests that have been performed with this and other games (Axelrod, 1984). It is important to appreciate, however, that this result is a consequence of learning and is not an intrinsic property of the model structure or of individual rationality. The players simply learn that, in the long run, it is in their better individual interest to cooperate. This result is relevant to the development of theories of learning and demonstrates how cooperation, and perhaps even notions of group preference, can evolve. Although such games serve as platforms with which to conduct important psychological experiments, the results do not alter the fact that the mathematical structure of categorical preferences and the logical structure of individual rationality do not accommodate an explicitly discernible notion of group preference and, hence, of group rationality.

One of the primary virtues of a mathematical model is that it provides a quantitative description of the values and preferences of the stakeholders. Ideally, such a model will strip the problem of all irrelevant and redundant issues and reduce it to its bare-bones mathematical essence. Rasmusen (1989) terms this no-fat modeling: "No-fat modeling is an extension of Occam's razor and the *ceteris paribus* assumption so fundamental to economics or, indeed, to any kind of analysis. The heart of the approach is to discover the simplest assumptions needed to generate an interesting conclusion: the starkest, barest, model that has the desired result" (pp. 14–15). This modeling assumption is compatible with the "hourglass" approach described by Slatkin (1980). According to Slatkin's approach, a complex problem is introduced, then reduced to a tractable mathematical model by stripping away all irrelevant issues, and finally, once a solution is obtained, expanded back into the original context for interpretation.

It is the last element of the hourglass approach: however, that is the most problematic. Many examples of social situations exist where classical game theory does an inconsistent job of explaining or predicting human behavior (e.g., the Prisoner's Dilemma, the Ultimatum game). Evidently, representing a social encounter by a set of categorical preference orderings can remove some meat along with the fat.

In an attempt to keep the meat on the bone, the field of behavioral economics has augmented the concept of self-interest to render it more psychologically realistic by incorporating notions of fairness and equity into the individuals' preference orderings. We illustrate this situation with the following example.

Example 1.2   The Ultimatum game has received great attention as a purported example of irrational behavior; that is, as a case where the players of the game are motivated by considerations other than maximizing their individual benefit. The setup of this two-player game is as follows: X_1, called the *proposer*, has access to a fortune, f, and offers X_2, called the *responder*, a portion $p \leq f$,

Table 1.2. *The payoff matrix for the Ultimatum game*

	X_2	
X_1	accept	reject
$p \in [0, f]$	$((f - p), p)$	$(0, 0)$

and X_2 chooses whether or not to accept the offer. If X_2 accepts, then the two players divide the fortune between themselves according to the agreed-upon portions, but if X_2 declines the offer, neither player receives anything. In both cases, the game is over; there is no opportunity for reconsideration. The payoff matrix for this game is illustrated in Table 1.2. The Nash equilibrium solution to this game is for X_1 to offer the smallest nonzero amount it can, and for X_2 to accept whatever is offered. This is the play predicted by individual rationality. Interestingly, such a strategy is rarely adopted by human players. Even with one-off play, proposers are inclined to offer fair deals ($p \approx f/2$), and responders are inclined to reject unfair deals ($p \approx 0$) (Binmore, 1998; Nowak et al., 2000; Sigmund et al., 2002). □

Although maximizing its payoff is the optimal choice for X_1, one can argue that it would be rational for X_1 to back off out of fear of losing everything, but that is a psychological consideration that is not part of the mathematical structure of the model under consideration. Another argument is that the players possess social attitudes, such as fairness, in addition to maximizing benefit. To this end, Fehr and Schmidt (1999) introduce the notion of "inequity aversion" into the game by adding a component to the preference model to account for the desire of the individual to avoid outcomes that strongly disadvantage any of the players. Once the preferences have been redefined, however, the game is played using the solution concepts that derive from individual rationality. As explained by Fehr and Schmidt (1999):

> We model fairness as self-centered inequity aversion. Inequity aversion means that people resist inequitable outcomes; i.e., they are willing to give up some material payoff to move in the direction of more equitable outcomes. *Inequity aversion is self-centered if people do not care per se about inequity that exists among other people but are only interested in the fairness of their own material payoff relative to the payoff of others* [emphasis added]. (p. 819)

Although this approach is certainly laudable, it does lead to some interesting issues. According to this view, possessing a notion of fairness does not imply that the player has any real concern for the welfare of others. On the contrary, the player acts in an unselfish manner simply because it dislikes inequitable outcomes – the effect on others is irrelevant. Such a player paradoxically has purely selfish reasons for acting unselfishly. Although necessary to retain allegiance to the doctrine of individual rationality, the somewhat tortured nature of this logic is unsettling. Simply relabeling fairness as "self-centered inequity aversion" is little more than a Procrustian-motivated semantic trick. Once the notion of fairness is invoked, it is both natural and appropriate to view behavior from the perspective of the interests of others as well as of the self. Once interests extend beyond the self, the notion of what constitutes rational behavior for the group can no longer be suppressed.

1.2.2 Rationality

One approach to defining rational behavior is to jump directly to a model of group interest that bypasses the notion of individual interest. The problem with such an approach is that it is difficult to reconcile it with individual autonomous action. Certainly, if a central authority could dictate the actions of all individuals, then the interests of the group would be served, but the interests of the individuals would be irrelevant. The only way a group preference could be defined under the individual rationality paradigm is if all stakeholders prefer the same outcome. In that situation, we may invoke the *Pareto principle*, also called the *principle of unanimity*, which posits that if all stakeholders prefer the same outcome, then the group prefers that outcome. In that ideal situation, at least, we may define a notion of rational behavior for the group from the point of view of individual rationality. But even if all players were to prefer the same outcome most, they would not necessarily agree on the next-most preferred outcome, and so forth; thus, a total group-level ordering of all outcomes is unlikely even in this case. Consequently, it is generally impossible to define a notion of group rationality if we restrict attention to, and only to, self-interest and categorical utilities. Luce and Raiffa (1957) put it this way: "General game theory seems to be in part a sociological theory which does not include any sociological assumptions, and, although one might hope one day to derive sociological theory from individual psychology, it may be too much to ask that any sociology be derived from the single assumption of individual rationality" (p. 196).

Nevertheless, the classical way to deal with social relationships is to assume that the processes by which preferences are defined is decoupled from the way the decision is made. Friedman (1961) puts it this way:

> The economist has little to say about the formation of wants; this is the province of the psychologist. The economist's task is to trace the consequences of any given set

of wants. The legitimacy of any justification for this abstraction must rest ultimately, in this case as with any other abstraction, on the light that is shed and the power to predict that is yielded by the abstraction. (p. 13)

According to this point of view, each stakeholder comes to the social encounter with a preference ordering that completely defines its personal tastes and values in a way that simultaneously accounts for such dissimilar motives as egoism, altruism, benevolence, malevolence, and indifference to the welfare of others, and that is not susceptible to change as a result of social interaction. This is a tall order, but nothing less will do if we are restricted to categorical preference orderings.

Simon (1986), however, criticizes the classical practice of eschewing concern for the processes by which individual preferences are generated.

Until very recently, neoclassical economics has developed no strong empirical methodology for investigating the processes whereby values are formed, for the content of the utility function lies outside its self-defined scope. It has developed no special methodology for investigating how particular aspects of reality, rather than other aspects, come to the decision maker's attention, or for investigating how a representation of the choice situation is formed, or for investigating how reasoning processes are applied to draw out the consequences of such representations. (p. 27)

Despite its limitations, individual rationality has served as the philosophical underpinning of many solution concepts. The mathematical structure of categorical preference orderings and the logical structure of individual rationality are ideally matched to each other. With these structures, it is possible to formulate precise definitions of what "best action" means in multistakeholder scenarios. This study has led to the vast literature of game theory, social choice theory, and multiagent systems theory. The key observation in this regard is the fact that, in all of these cases, the focus is on justifying the choices from the point of view of the individual, with less attention given to considering rationality from the point of view of the group.

In a mixed-motive context, however, deciding what is categorically "best" may be difficult to ascertain. If social relationships exist among the players, then what is good for one may depend on what is good for others, and the individual must somehow strike a balance between its interests and the interests of others. Consequently, a more general concept of rational behavior is desirable. To identify such a concept, we begin by considering two distinct notions of rationality, as identified by Simon (1986): one that focuses on justification and one that focuses on methods. The former he terms *substantive rationality*, which is the doctrine that the reasons why a decision is made are of paramount importance, and the latter he terms *procedural rationality*, which is the doctrine that the methods used to arrive at a decision are most important. Substantive rationality is prescriptive – it tells us where to go but not how to get there – whereas procedural rationality is descriptive – it tells us how to get there, but not where to go. The former focuses on the result; the latter focuses on the process.

Individual rationality is the prime example of the substantively rational approach to decision making. It is a very limited concept and does not easily accommodate notions such as reciprocity, generosity, malevolence, and benevolence, particularly when exercising these attributes can negatively impact one's own welfare. Our objective is to expand the scope of the currently accepted foundational elements of decision theory in a way that retains as much simplicity as possible but does not surrender an adequate representation of relevant behavior. Essentially, we seek middle ground between the substantive and procedural notions of rationality that accommodates a mathematically rigorous concept of quality assessment (the substantive component) and accounts for the way participants can influence each other's preferences (the procedural component). We focus on situations where the relationships among the stakeholders are too complex to express in terms of categorical preferences, and where notions of self-interest are not fully compatible with desired or observed behavior. In such situations, a more sophisticated framework that accounts for social relationships and relies on the notions of rationality that such relationships can engender is required.[5]

To appreciate the importance of reasoning processes, let us consider an example, adapted from Keeney and Raiffa (1993) of a farmer and his wife who must decide among four different climates: much-rain/much-sun, much-rain/little-sun, little-rain/much-sun, and little-rain/little-sun. Suppose the farmer's wife is concerned, perhaps for health reasons, about rainfall, while the farmer is concerned about the success of his crop, and therefore must consider both rainfall and sunshine. He is also concerned about the welfare of his wife. Thus, his preference ordering for sunshine if she were to prefer a much-rain climate could be different from his preference ordering for sunshine if she were to prefer a little-rain climate.

The farmer does not come to the decision problem with a definitive preference ordering in place and must wait until he interacts with his wife to resolve his ambiguous preferences. It may be argued, however, that once this is done, there is no real difference between this scenario and the conventional scenario where all individuals come to the decision problem with categorical preferences. In other words, once the psychological influence that his wife exerts on the farmer takes its effect, Friedman's (1961) division of labor is faithfully realized. But

---

[5] One is tempted to call this concept "rational socialism."

once we know that the farmer has ambivalent preferences, we must be assured that he resolves his ambivalence in a rational way. Elster (1986) argues that an action cannot be considered to be rational unless it can be justified.

> Once we have constructed a normative theory of rational choice, we may go on to employ it for explanatory purposes. We may ask, that is, whether a given action was performed *because* it was rational. To show that it was, it is not sufficient to show that the action was rational, since people are sometimes lead by accident or coincidence to do what is in fact best for them. We must show, in addition, that the action arose in the proper way, through a proper kind of connection to desires, beliefs, and evidence [emphasis in original]. (p. 2)

If the farmer were to come to the decision problem with unique categorical preferences and used them to make a choice, we might be interested in his reasoning process that lead him to those preferences, but it would not be reasonable for us to require him to defend his preference ordering before we would be willing to grant that his decision has a legitimate claim to rationality. But if the farmer were to come to the decision problem while holding his preferences in suspense pending knowledge of his wife's preferences, we cannot know that the decision is rational unless we understand the *process* by which the ambiguity is resolved. Even if he happened to make the best choice for both his wife and himself, the way the farmer settled on his choice could not be ignored.

Nevertheless, the view that the process by which preferences are formulated is irrelevant to the decision problem dominates much of decision theory. Many theorists simply adopt the Arrow–Friedman blanket assumption and use the existence of individual preferences as a starting point. This position is consistent with the principle of *methodological individualism* – the doctrine that the behavior of a society must be explained by showing how the behavior results from the intentional states that motivate its individual members (Weber, 1968; Homans, 1967). But methodological individualism does not insist that all individuals come to the decision problem with definitive preference orderings. Nor does it prohibit participants from modifying their preferences as a result of interacting with others. Although the farmer may not possess a unique preference ordering before he interacts with his wife, the choice facing him and his wife certainly can be made on the basis of the intentional states that motivate the members. All that is required is that the intentional state of the farmer depends on the possible intentional states of his wife. He cannot resolve his ambiguity without considering her possible intentional states.

1.2.3 Influence

Unless an individual stakeholder is completely oblivious to the behavior of others, the possible choices of others will influence its choices (indeed, this assumption provides the principle rationale for the Nash equilibrium concept). Under the regime of categorical preferences, however, such influence is *tacit*, in that it is implied through the solution concepts that are employed by the individual, but is not explicitly present in the payoffs. With solution concepts based on individual rationality, such influence is restricted to considerations of what other stakeholders can do, but *not with respect to considerations of what they prefer to do.*

Once we consider scenarios where stakeholders possess more than a minimal concept of sociality, we may consider the possibility that a stakeholder maybe influenced by concerns for not only what actions other stakeholders can perform but also what outcomes they prefer. Sen (1979b) argues that an individual who is part of a group is influenced by social relationships, and those relationships cause the individual to expand its sphere of interest beyond the self and consider the interests of others.

> It is possible to argue that a theory of group choice should be concerned merely with the derivation of social preference from a set of individual preferences, and need not go into the formation of individual preferences themselves. This view has its attractions, not the least of which is its convenience in limiting the exercise. However, it is a somewhat narrow position to take. ... The society in which a person lives, the class to which he belongs, the relation that he has with the social and economic structure of the community, are all relevant to a person's choice not merely because they affect the nature of his personal interests but also because they influence his value system including his notion of "due" concern for other members of society. (pp. 5–6)

Although consideration of the process by which preferences are formed may not be relevant when competition is the dominant feature of a group, this argument loses much of its power for mixed-motive multistakeholder decision problems. It may be difficult, and perhaps impossible, for an individual to express its preferences simultaneously as both categorical and fully representative of its social considerations. An alternative is for the stakeholder to express its preferences with a more complex framework. One way to do this is to subscribe to the argument espoused by Goodin (1986) that stakeholders may possess metapreferences; that is, preferences for preferences.

> On some accounts, people are distinguished from lower forms of life precisely by the fact that they have not only preferences but also preferences for preferences (Frankfurt, 1971). These may be moral ideals (Sen, 1979a), personal ideals, social ideals or role preferences (Goodin, 1975; Benn, 1976, 1979; Hollis, 1977). ... Frequently, such individuals find that one set of preferences actually guides their behavior while they dearly wish another would instead. Laundering their preferences then simply amounts to respecting their own preferences for

preferences. In aggregating preferences, we count only those the individual wishes he had; and we ignore all those he wishes he did not actually experience. (p. 83)

Distilling dissimilar individual preferences into a single categorical ordering can strip the representation of important, and even essential, social content. Furthermore, categorical preferences do not permit the opportunity for stakeholders to modify their preferences upon contact with others. Goodin (1986) argues that such modifications are essential.

> Formal models of collective choice tend to represent it as some mechanical process of aggregating individual preferences. This badly understates the true complexity of the process. Whereas these models usually take preferences as given, for example, classical theories of democracy have always acknowledged that people can and should reformulate their preferences in response to rational discussions in the course of collective deliberations: instead of working on some *fixed* set of preferences, the social decision machinery changes them in the process of aggregating them [emphasis in original]. (pp. 86–87)

Axelrod (1984) lists four factors that give rise to complex social structures that go beyond narrow self-interest, which he terms labels, reputation, regulation, and territoriality. These factors may motivate a player to modify its strategy to take advantage of other players' social propensities, even to the point of overriding its ex ante preference orderings based solely on payoffs. For example, Axelrod (1984) argues that a player of the Prisoner's Dilemma game who has a reputation for being "nice" can elicit a response from other players that is very different from the response dictated by strictly "rational" logic. If a player knows or has learned the social traits of other players, it can make strategic alterations to its behavior with confidence. However, if the player does not have complete confidence in its assessment of the social propensities of others, it may nevertheless generate a family of contingency strategies, with each one based on a different hypothesis regarding the other players' social traits. In fact, the player can even do more. It can move beyond a merely tacit assessment of other players' motives and systematically modulate its preference orderings based on the different social hypotheses.

A natural mechanism for such modulation is the notion of influence. Let X_1 and X_2 be two stakeholders. We say that X_1 *socially influences* X_2 if X_2's preferences are affected by X_1's preferences such that without knowledge of X_1's preferences, X_2 is in a state of suspension with respect to its own preferences. Thus, X_2 possesses multiple preference orderings and cannot choose from among them without information regarding X_1's preferences. In other words, X_2 must define a preference ordering for its preferences. Let us assume, however, that once X_1's preferences are revealed, X_2 will no longer be in suspense and can choose an appropriate preference ordering for itself. Essentially,

X_1's preferences propagate through the group to affect X_2's preferences, thereby generating a social bond between the stakeholders.

The existence of social influence and the generation of social bonds is a manifestation of expanding beyond the strict confines of self-interest and presents the possibility that players may rationally behave in ways that are not strictly aligned with their payoffs. To see how this might happen, let us reconsider the Prisoner's Dilemma. While keeping the payoff matrix as given in Table 1.1, let us suppose that from a social point of view, the critical issues are (a) the propensity of an individual to cooperate and (b) the propensity of an individual to exploit. A player who possesses a strong propensity to cooperate may be willing to forgo the payoff of exploitation and even risk being exploited itself in an attempt to cooperate. For example, if X_1 were to have a higher propensity to cooperate than to exploit, then it would prefer (C,C) to (D,C); that is, X_1 would prefer its individually next-best outcome to its best outcome.

Let us also suppose that X_2 possesses these social attributes, although not necessarily to the same degree as X_1. It would be reasonable for X_2 to consider its preferences in the light of X_1's preferences. Under the hypothesis that X_1 prefers (D,C) to (C,C), X_2 would most likely prefer (C,D) to (C,C), but under the hypothesis that X_1 prefers (C,C) to (D,C), X_2 might well prefer (C,C) to (C,D) (depending on its own social propensities). Thus, X_2 possesses two preference orderings: one that corresponds to the hypothesis that X_1 is inclined to be cooperative, and the other that correspond to the hypothesis that X_1 is inclined to be exploitive. Let us term preference orderings of this type *conditional preference orderings*.

If both players were to prefer mutual cooperation to either exploitation or mutual defection, then the Pareto principle would apply, and the group would prefer mutual cooperation to every other outcome. Thus, although the game would still be a PD in terms of actual payoffs as defined by Table 1.1, the social propensities possessed by the players could result in an outcome different from the Nash equilibrium solution of mutual defection.

One may object to relaxing the requirement for strict alignment of preferences and payoffs. It may be argued that introducing social attributes and letting them govern the decision rule to the extent of overriding the payoff-defined ordering is inappropriate since it can lead to inconsistency between the individuals' subjective social valuations and their objective economic valuations. Is not this situation merely evidence that the individual has not yet completely defined either its preferences or its payoffs, and should be sent back to the drawing board? Perhaps so. But it also may be true that such inconsistencies are real, and attempting to eliminate them only leads to inconsistencies popping up in other places, such as requiring purely selfish reasons for acting unselfishly.

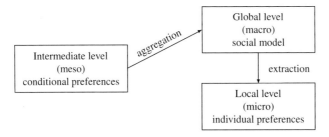

Figure 1.1. Flow of social influence.

Chapter 2 presents a new mathematical structure to define conditional preferences that link individuals together in a way that it is difficult to duplicate
with categorical preferences. The basic idea is to define influence flows between
individuals in the form of conditional preferences and to allow these influence
flows to propagate through the entire group, as illustrated by Figure 1.1. This
propagation constitutes a meso-to-macro aggregation from the intermediate, or
meso, level of conditional preferences to an emergent global, or macro, level
notion of group preference, which, although not in a form to define group-level
action, provides an ordering of the degree of conflict and harmony that exists
among the members of the group. This ordering, which we term *concordance*,
provides a model of the social structure of the group. Using this global social
model, we define a macro-to-micro procedure to extract individual preference
orderings that take into account all social relationships that exist among the
members of the group. These micro-level preference orderings may then be
used to define rational behavior for the individuals that extends beyond narrow
self-interest.

Although the group, as an entity, is not empowered to take collective action, it
is nevertheless possible for individuals to define and take into consideration the
concordance of the group when making their individual choices. In Chapter 3 we
introduce new solution concepts that establish a basis for negotiation whereby
the members of a group strive to achieve an acceptable outcome for the group as
well as for themselves. Since these solution concepts involve conditional preference orderings, they can be more conceptually and computationally complex
than solution concepts that involve only categorical preference orderings. But
as we shall see, there are often ways to mitigate the complexity in ways that
still lead to conceptually and computationally tractable models. Regardless
of the possible increase in complexity, however, it cannot be avoided if the
model is to be an appropriate social characterization of the group. As Palmer
(1971) observes, "complexity is no argument against a theoretical approach if

the complexity arises not out of the theory itself but out of the material which any theory ought to handle" (p. 176).

1.3 Coordination

Once the interests of the stakeholders extend beyond the self, it is possible to evaluate the ability of the group to coordinate. According to the *Oxford English Dictionary*, *to coordinate* is "to place or arrange (things) in proper position relative to each other and to the system of which they form parts; to bring into proper combined order as parts of a whole" (Murray et al., 1991). In the context of a multistakeholder group, coordination means that behavior of the individuals (the parts) must be reconciled with behavior of the group (the whole). To satisfy the dictionary definition, therefore, true coordination requires consideration of the proper behavior of the group as well as of its members.

Although the terms are sometimes conflated, coordination is not the same thing as cooperation. While agents may cooperate to the individual advantage of each, such as when their individual interests just happen to be compatible, that does not necessarily mean that a concept of preference for the group as a whole exists or can be defined. With the Prisoner's Dilemma, for example, if both players choose to cooperate, they both receive a better outcome than if both choose to defect, but those are strictly individual benefits. Of course, it is always possible for an outside entity to impose a notion of group preference, but no such notion follows as a logical consequence of individual categorical preferences.

Game-theoretic treatments of coordination, as discussed by Schelling (1960), Lewis (1969), Bicchieri (1993), Malone and Crowston (1991, 2003), Cooper (1999), and Goyal (2007), are based upon the classical notion of categorical preference orderings. When the payoffs of the players are juxtaposed in a payoff array, the opportunities for exploitation, compromise, altruism, and cooperation become apparent. It is assumed that the propensities of the individuals with respect to any of these modes of behavior have been thoroughly considered by each player and have been encoded into its payoff. Once the payoffs have been defined, however, the rationale behind their generation is, according to Friedman (1961), irrelevant to the solution. Unfortunately, however, it is exactly these behavioral attributes that come into play when seeking a "coordinated" solution. The concept of "tacit coordination" is defined by Schelling (1960): "in tacit coordination . . . one is trying to guess what the other will guess one's self to guess the other to guess, and so on ad infinitum" (pp. 92–93). Schelling goes on to observe that "In the pure-coordination game, the player's objective is to make contact with the other player through some imaginative process of introspection, of searching for shared clues" (p. 96).

Tacit coordination requires the players to know things about each other in addition to the information contained in the payoff array. It assumes that there are sociological and psychological issues that influence the attitudes and motives in ways that are not expressed with the categorical preferences. In fact, Bicchieri (1993) argues that coordination is unlikely unless the players have significant social knowledge about each other, either known a priori or obtained as a result of learning. Unless such information is encoded in the payoffs, however, it is not part of the formal structure of the game. Coordination that derives from the extra-game knowledge that the players have regarding each other is termed *extrinsic coordination*, since it is derived from considerations that are not part of the formal structure of the game as defined by the payoffs.

Example 1.3 A classical example of a game where extrinsic coordination is essential is the Battle of the Sexes, a well-known, non-zero-sum, two-player game involving a man and a woman who plan to meet in town for a social function. She (S) prefers to go to the ballet (B), while he (H) prefers the dog races (D). Each prefers to be with the other, however, regardless of where the social function takes place. The payoff matrix in ordinal form is given in Table 1.3. This game has two Nash equilibria, (D, D) and (B, B). The storyline for the game presents a natural environment for the incorporation of sociological issues, and it may be argued that the standard model, as defined by the payoffs, is inadequate, since only tacit coordination is possible. Of course, from a purely theoretical point of view, the storyline is simply a convenient setting to motivate the payoff structure. Nevertheless, if the game is to be of any practical use as it relates to the storyline, it must withstand sociological scrutiny. Indeed, the payoff matrix might be more appropriate if the two players were complete strangers, in which case, opportunities for tacit coordination would be diminished. □

The ability to coordinate extrinsically depends upon the propensities, introspective skills, and concepts of rationality employed by the players as they attempt to synthesize a notion of group preference. This synthesis may lead to the effective playing of the game, but the extrinsic approach does not address the issue of how amenable the game itself is, in terms of its preference structure, to combine the parts properly to form a whole. This concern leads us to consider a concept of *intrinsic coordination* to characterize the innate ability of a group to coordinate. Intrinsic coordination arises as a result of the social influences that exist among the members of a group as objectively encoded in the preference relationships, rather than from the subjective psychological

Table 1.3. *The payoff matrix in ordinal form for the Battle of the Sexes game*

	S	
H	*D*	*B*
D	(4, 3)	(2, 2)
B	(1, 1)	(3, 4)

Key: 4 = best; 3 = next-best; 2 = next-worst; 1 = worst

attitudes and rationality concepts of the players that are supplied by external observers. If all members possess categorical preferences, then no notion of group preference exists and, consequently, they possess no intrinsic ability to coordinate (although they may cooperate if their payoffs are compatible). However, if the members are able to influence each other's preferences, then a notion of group preference may emerge, and the members may be able to coordinate.

Assessing intrinsic coordination requires an investigation of the structural properties of the preference orderings as social influence propagates through the group. This study must be in greater depth than simply an investigation of the opportunities for cooperation or competition that become apparent once the payoffs of the players are juxtaposed in a payoff array. In this treatment, coordination is as a neutral concept – neither necessarily positive, such as cooperation, nor negative, such as competition. We adopt the dictionary definition and require coordination to involve some form of sociality among the stakeholders that corresponds to group behavior. Essentially, we follow the dictum espoused by Ortega y Gasset (1975): "Order is not pressure which is imposed on society from without, but an equilibrium which is set up from within." Our objective is to develop a quantitative measure of the ability to reconcile the behavior of the individuals with the desired behavior of the group. Such information would serve as metaknowledge that would indicate the intrinsic ability of the members of a group to function harmoniously, regardless of their notions of rationality. As will be developed in Chapter 4, the creation of a quantitative measure of the ability to coordinate will draw heavily on concepts of entropy and mutual information, as developed by Shannon and Weaver (1949).

1.4 Uncertainty

No discussion of rational decision making would be complete without addressing the issue of how to make decisions when there is a lack of complete knowledge. This lack necessitates a fundamental change in perspective, as articulated by Jeffyrey (1992):

> Outside the realm of what we are sure of lies the puzzling region of probable knowledge – puzzling in part because the sense of the noun seems to be canceled by that of the adjective. The obvious move is to deny that the notion of knowledge has the importance generally attributed to it, and to try to make the concept of belief do the work that philosophers have generally assigned the grander concept. I shall argue that this is the right move. (p. 30)

This change in perspective from making decisions on the basis of knowledge to making them on the basis of belief requires that, in addition to defining mathematical models to characterize preferences, we must also define mathematical models to characterize belief.

Probability theory is by far the best-known and most heavily used mathematical model with which to characterize uncertainty, and the most widely used notion of rational behavior is to make choices on the basis of expected utility – that is, the sum of the products of the utilities of the outcomes and their respective probabilities of instantiation. Expected utility represents the valuation point such that the stakeholder is indifferent between receiving that value for certain and receiving the value corresponding to the actual outcome (which is unknown prior to making a choice). Accordingly, the operative notion of individually rational decision making in the presence of uncertainty is to maximize expected utility.

There are many ways by which uncertainty can enter into a multistakeholder decision problem, but we will restrict attention to those cases where the sources of the uncertainty can be modeled as stakeholders whose choices are governed by probability distributions. Thus, the group may comprise two sets of stakeholders: deterministic stakeholders whose actions are governed by their preference orderings and stochastic stakeholders whose actions are governed by probability distributions. We also assume that these two sets of stakeholders can influence each other. Chapter 5 illustrates how the probabilistic model for the stochastic stakeholders' behavior can be merged with the conditional preference model for the deterministic stakeholders' behavior to provide a unified treatment whereby the behavior of both groups of stakeholders can be represented by a single mathematical structure.

1.5 Satisficing

The solution concepts of classical decision theory and classical game theory are designed to produce a choice that is "best" according to some relevant notion of value. The notion of being "best," however, is a limiting concept of acceptability in a multistakeholder context. It depends on a rigid point of view that is consistent with, and only with, the notions of categorical preferences and individual rationality. Since the thesis of this book is to generalize both of those notions, it makes sense also to generalize the notion of what it means to be acceptable. An alternative notion of rational behavior that is more flexible than demanding the best and only the best outcome is to focus on outcomes that are "good enough" in some well-defined sense. What is best for me may not be best for you, but there may be some outcome that is good enough for both of us. Chapter 6 presents a formal mathematical definition for the concept of being good enough, or *satisficing*. As used here, however, the notion of satisficing, although similar to Simon's (1955) usage in the sense that it identifies solutions that, while perhaps not optimal, are good enough, it differs in that, whereas Simon's approach is essentially heuristic and is expressed in terms of subjectively defined aspiration levels, the methodology presented herein employs a mathematical definition that is as rigorous in structure as is the concept of optimization. Chapter 6 then develops satisficing solution concepts for groups and individuals.

1.6 Applications

To motivate the need to reconsider the foundations of classical theories of multistakeholder decision theory, we provide a brief summary review of classical approaches, as applied to traditional economic and sociological contexts, where categorical preference orderings are currently almost always used. We then consider the relatively new field of artificially intelligent societies and discuss some of the important issues regarding the design and synthesis of such systems.

1.6.1 Classical game theory

As established in their seminal book, von Neumann and Morgenstern (1944) built a theory of multistakeholder decision making on the fundamental premise that the players are autonomous and individually possess the ability to take action. There is no central authority that dictates the behavior of the players. If such an authority were to exist, the game could be viewed as an unconstrained n-player optimization problem, where the optimality criterion is defined by the authority. Since that is not the case, the solution becomes an exercise in solving *n* one-player constrained optimization problems, where the criteria for optimality are defined by the players.

There are two major classes of games: so-called noncooperative games and cooperative games. With a noncooperative game, each player operates independently, while with a cooperative game, the players are able to form binding agreements, or coalitions. This terminology can be misleading. Players of a noncooperative game may indeed behave cooperatively if it is in their individually best interests to do so, while players of a cooperative game may be fiercely competitive. In both cases, all players act in accordance with their self-interest.

In strategic form, a noncooperative single-stage game is represented by a payoff array, which is formed by juxtaposing the individual payoffs. Once so juxtaposed, opportunities for cooperation and conflict become apparent, and some solution concept, such as Nash equilibrium, may be applied to define a solution. Nash equilibrium, however, is a manifestation of individual rationality. It constitutes a simultaneous constrained maximization problem where all stakeholders operate under the same principle: identify a solution such that if any single player, acting alone, were to deviate, the benefit to that player would decrease.

By its name, it might be supposed that cooperative game theory possesses some notion of group rationality. A subset of players who agree on their strategies can receive a joint payoff. If the joint payoff is superadditive and utility is transferable, then the payoff to the subset is greater than the sum of the individual payoffs, and each member of the coalition can be assured of receiving a benefit that is greater than it would receive if it did not join the coalition. There are a number of solution concepts for cooperative games, including the von Neumann–Morgenstern solution, the core, the Shapely value, and the Nash bargain. At the end of the day, however, cooperative game theory is also based squarely upon the assumption of individual rationality. Each player enters into a coalition solely on the basis of benefit to itself, and even though each may be better off for having joined, a notion of group benefit is not an issue when forming the coalition, nor does such a notion even need to be defined in an operational way.

One of the recent advances in game theory is the postulation of evolutionary processes to explain cooperation as a result of players mutating over successive generations and reproducing themselves according to notions of fitness for survival (Maynard-Smith, 1982; Axelrod and Hamilton, 1981; Weibull, 1995). Such evolutionary models are used to demonstrate that players are not exclusively motivated by narrow self-interest but also care about the payoffs and intentions of others (Fehr and Gächter, 2002; Gneezy, 2005). While these are important advances, they focus mainly on alternative notions of rational behavior, concepts of fitness, and biological and social evolutionary processes. Evolutionary game theory provides an important model of social relationship emergence as games are played in environments where players' fitness for long-term survival is taken into consideration in addition to their short-term payoffs. It is important to appreciate, however, that the additional notions of fitness and replication are extrinsic to the game's mathematical structure. Essentially, these notions merely overlay the same categorical utility and individual rationality structures that are employed by classical theory.

Although classical game theory (either noncooperative or cooperative) is a powerful mathematical tool for the analysis of multistakeholder decision making, reliance on categorical utilities limits its usefulness to situations where self-interest is the appropriate social model. In that context, the various solution concepts, such as Nash equilibrium, describe behavior that can be interpreted as rational. Even then, however, Simon (1996) offers the following observation: "Game theory's most valuable contribution has been to show that rationality is effectively undefinable when competitive actors have unlimited computational capabilities for outguessing each other" (p. 38). Notwithstanding its limitations, however, game theory is well established as a valuable tool for the study of economic systems. More recently, it has been used for engineering and computer science applications such as mechanism design, power system design, and network design.

1.6.2 Classical social choice theory

Social choice theory is, essentially, a theory of democratic decision making in the sense that the decision made by a group should be based on the preferences of the individuals who compose the group. In contrast to the classical game theory model which allows each stakeholder to make its own decision, the social choice model consists of a set of alternatives, of which one, and only one, can be implemented by the group, which comprises two or more stakeholders, or voters.

An essential difference between social choice theory and game theory is the ability of the stakeholders to implement their choices. With game theory, each stakeholder is empowered to implement its preferred choice, although the outcome depends on the choices of the other players as well. The stakeholders in the social choice context, however, are not empowered to enforce their preferred choice. Rather, their preferences are combined (for example, via a voting mechanism or some aggregation procedure) to arrive at a single choice for the entire group. Essentially, game theory is a model of every-man-for-himself, with group performance taking its chances, while social choice is a model of making a choice for the group, with individuals taking their chances. Thus, the role of a stakeholder in a social choice scenario is to express its preference and then to submit to the choice of the group.

One of the fundamental theoretical results associated with social choice theory is Arrow's (1951) celebrated impossibility theorem: It is generally impossible to aggregate a set of individual preference orderings to form a preference ordering for the group and still comply with a set of desirable conditions (unrestricted domain, the weak Pareto principle, the independence of irrelevant alternatives, and nondictatorship). It is important to appreciate that Arrow's result applies to ordinal binary preference relations that constitute a total ordering of the alternative set. As Shubik (1982) has observed, however, if the individual orderings are expressed via real-valued utilities, then it is possible to define a numerical ordering for the group that does indeed satisfy the previously-mentioned conditions. For example, it is easily seen that the sum of the individual utilities satisfies the conditions. In fact, there is an infinity of such functions. Thus, in terms of binary relations, there are too few ways to aggregate, but two many ways to aggregate in terms of numerical orderings.

The distinction between the two formulations of aggregation is subtle, but important, and, as Shubik (1982) contends, often not sufficiently emphasized by social choice theorists.

> The very existence of the [numerical formulation] shows that the "impossibility of a social welfare function" is not a robust principle. It is a theorem pertaining to a particular formal system. In order to carry this theorem over to real societies, one must establish that the distinctive conditions of the formal system are met – in particular, that the individuals in the real situation are not only ordinalists but relativists, able to discern better from worse but not good from bad. (p. 122)

Similar to classical game theory, classical social choice theory begins with the assumption that all stakeholders come to the issue with categorical preference orderings. This point is emphasized by Johnson (1998).

> In social choice theory, as in the broader field of rational choice, individual goals are typically taken as "givens," part of the data provided by a study of a particular situation. This is practical decision, based in large part on the need to keep research projects manageable. Asking why people like the things they like or why they have a particular political ideology would take us too far from the focal point, which is preference aggregation. (p. 4)

Thus, classical social choice theory adheres to Friedman's (1961) division of labor hypothesis. It is possible, however, to focus at a deeper level and argue that an understanding of the relationship between individual preferences and group decisions requires an understanding of the influence relationships that exist among the individuals. The following scenarios illustrate the demand for such a deeper understanding:

- **Elections**. Elections involve the combination of individual views of who should be appointed to some office. Given a finite set of candidates, the simplest and most common model is for each individual X_i to possess its own ranking of the candidates. In many societies, however, although all individuals have the same electing power (one vote each), the views of the electorate are not formed in a social vacuum. That is, it is generally not the case that each individual considers, in isolation from others, his or her evaluations of the candidates. Often, some electors are able to influence others, positively or negatively. For example, suppose candidates A and B are up for election. Mary, a respected member of the community, announces that she prefers candidate A. Those people whom Mary influences may be swayed – not by of the merits of the candidate, but by Mary's preference. The preferences of those who are so influenced are thus dependent upon the preferences of Mary. This influence may be positive (they support candidate A if Mary does) or negative (they support B, whom Mary opposes). However, under the Arrow–Friedman regime, the way in which these individuals come to their preferences has no bearing on the choice that is made. Yet, it seems that the fact that some members of a society are able to influence others ought not to be ignored.
- **Marketing**. Consumer preferences influence the availability of products in the market. But consumers do not determine their preferences in social isolation. Typically, they are influenced by the opinions of others as well as by advertising, price, and quality. Thus, to suppose that each individual consumer somehow arrives at its own preference ordering, but the way that ordering is determined is irrelevant, is overly simplistic, if not naive. From the point of view of the suppliers and marketers of the commodity, it seems that the way in which the consumers arrive at their preferences ought to be important if they want to influence those preferences.
- **Moral Judgment**. It can be the case that an individual's personal interests conflict with those of others. One may disapprove of extremely violent films while respecting the right of others to choose for themselves whether or not to view them. Thus, it may turn out that an individual may lend support to some social proposition without personally supporting that proposition. When offering such support, it is essential to control the degree of support as a function of the strength of one's personal interests and the strength of one's concern for the interests of others.

1.6.3 Artificially intelligent societies

Game theory and social choice theory traditionally have dealt with human societies and, therefore, have focused mainly on political, economic, sociological, and psychological applications. Distributed control and multiagent systems problems represent a hybrid that includes both game-theoretic and social choice-theoretic components. Similar to game theory, each agent (or controller) possesses a preference ordering as a function of the choices of all agents. In addition to concern for the interests of the individual, some notion of group-level performance is also important. This dual nature creates a dichotomy: It is necessary to honor each stakeholder's interest while at the same time considering the interests of the group.

Often, a multiagent system is populated with individual agents who are designed to coordinate with each other to achieve some task or goal for themselves and for the group. In the absence of a centralized authority, however, the individual agents must each choose actions that satisfy any selfish notions of behavior (for example, energy conservation, exposure to hazard, violation of constraints) as well as meet the objectives of the group. Seeking an optimal solution is too simple, since a notion of optimal coordination may not be easily defined – again, it depends upon a particular point of view. Optimization, by its very nature, is an individual decision, so if a group is to optimize, it must act as if it were a single individual. Consequently, more complex notions of behavior are essential to the success of a distributed multiagent system.

The following list provides a sample of the possible applications of multiagent systems:

• An ad hoc wireless communications network involves a number of individuals who must communicate with each other. Typically, such networks are power-limited, and communication requires that messages be relayed through intermediate nodes (agents) of the system. On the one hand, relaying consumes energy and bandwidth, while on the other hand, an agent may need the help of others to be successful. Thus, the agents can be conflicted. Each must find a balance between its need to conserve resources and the necessity of establishing a cooperative reputation. The interactions between agents play an essential role in the ability of the network to be functional.

• Job shop scheduling and management problems involve a set of jobs and a set of machines. Each machine can handle one job at a time, with each job consisting of a chain of operations that must be processed in a given time frame. The problem is to allocate the operations to time intervals of minimum duration required to complete the job. Although, in theory, a centralized controller could solve this problem, unforeseen delays, equipment malfunctions, and other sources of random behavior suggest that a more robust approach would require local control, with the agents (the machines) scheduling via negotiation. This procedure would permit the tradeoff between local performance and global performance, provide higher fault tolerance, permit rescheduling, and so forth.

- Mobile robots can be used for surveillance, search and rescue, hazardous waste control, and other such applications. Such a system of distributed autonomous agents moving in a geographic environment must coordinate in order to search for targets of interest and avoid obstacles and threats. They are usually equipped with sensors to view a limited region of the environment, and they are able to communicate with other agents. They must comply with physical constraints such as maneuverability limitations, fuel and time constraints, avoidance of hazards, and sensor range and accuracy. The success of such a system depends on the robots' ability to coordinate in order to avoid duplication of effort, ensure full coverage of the geographic area, and comply with constraints such as maintaining communication links.

- Future generations of air traffic control management will provide pilots of appropriately equipped aircraft the freedom to manage their flight paths in real time. In such an environment, pilots will be given increased autonomy to select or modify their flight paths to avoid collisions, conserve fuel, and maintain time schedules. To do so will require the aircraft to communicate with each other and to arrive at negotiated decisions.

1.7 Summary

The development of theories for making rational decisions in multistakeholder environments is one of the important contributions of twentieth-century mathematics. The most influential contributions to these theories are von Neumann's game theory (von Neumann and Morgenstern, 1944) and Arrow's social choice theory (Arrow, 1951). The starting point for these theories, as well as virtually all subsequent embellishments and extensions of the original developments, is the two-part assumption that (a) each stakeholder comes to the decision problem with preferences that are completely and unconditionally defined before the individuals engage in the decision-making enterprise, and (b) each stakeholder's interest is limited to its own welfare. These assumptions are tantamount to viewing the stakeholders as completely independent entities with no social relationships, which justifies focusing on individual rationality as the appropriate rationality doctrine with which to define solutions.

If each stakeholder in every decision problem were to possess a categorical preference ordering that provided a complete definition of its preferences over all possible outcomes, and if each were always motivated exclusively by the individual rationality doctrine (however self-interest is defined), there would be no need to write this book, since the approaches of classical game theory, classical social choice theory, and conventional multiagent system theory have been developed to a high degree and have been proved to be successful in such circumstances. But when a stakeholder's sphere of interest extends beyond itself, the stakeholder may desire to consider the preferences of others when defining its own preferences, and it may not be convenient, straightforward, or even possible to define its preferences categorically. Thus, it is imperative that we investigate ways to frame multistakeholder decision problems that naturally permit such extensions of interest. We must also look beyond the individual rationality doctrine as the only operative concept of rational behavior.

Relying exclusively on categorical preference orderings strips game theory of the need to consider the processes by which preferences are formed. In a social context, however, decision making is a process. The players must interact, and the game must actually be played. Certainly, abstracting a decision scenario to a mathematical structure comprising players, actions, outcomes, and preferences is a powerful and even necessary step in order to quantify the game so that rationality concepts can be applied, but it is possible to overdo the abstracting. When social influences exist, it is essential that the processes by which they affect preferences be included explicitly in the mathematical model and not merely viewed as psychological issues that can subjectively influence the way the game is played, especially when the observed play is not in accordance with the presumed notions of rationality. In a nutshell, when social influences exist, they must be part of the mathematical model.

This book makes no pretense of explaining how humans actually make decisions. That remains the purview of the social and behavioral sciences. Although this extended model may provide a more flexible approximation of human behavior than conventional models offer, it is nevertheless a mathematical abstraction developed in the same spirit as is the model of classical game theory. Ultimately, we must concur with Friedman (1961) that the legitimacy of such an abstraction rests on the light that it sheds.

2

Conditioning

The manner in which mathematical theories are applied does not depend
on preconceived ideas: it is a purposeful technique depending on, and
changing with, experience.

    — *William Feller*
    An Introduction to Probability Theory and Its
    Applications (Wiley, 1950)

Given a group of stakeholders, a classical rational decision-making model comprises three distinct structural elements. First is the set of feasible actions (those actions that satisfy the logical, physical, and economic constraints associated with each stakeholder); second is the set of possible outcomes that can obtain as a result of all players taking action; and third is a preference ordering of the outcomes for each stakeholder. There is also a fourth component, namely, the concept of logic, or rationality, that governs decision making, but that component is not a part of the model structure. In this chapter, we focus exclusively on the structural components and defer consideration of rationality until Chapter 3.

As discussed in Chapter 1, under the classical game theory model, all players come to the moment of decision with all of their preference orderings completely and categorically defined. This model, however, does not permit group-level preferences to be defined, which significantly limits the use of classical game theory as a model of groups whose members possess sophisticated social relationships. This limitation is addressed by replacing categorical preference orderings with conditional preference orderings that explicitly account for the social influence that the players exert on each other. As the players interact, the influence propagates throughout the group, thereby creating a social bond that connects the interests of the players to each other and leads to a well-defined social model for the group. The goal of this chapter is to construct a mathematical description of this process.

We set the following notation: Let $\mathcal{X}_n = \{X_1,\ldots,X_n\}$, $n \geq 2$, denote a group of stakeholders, with X_i referring to the ith member of $\mathcal{X}_n$. Let $\mathcal{A}_i$ denote the space of feasible actions available to X_i, and let $a_i, a_i', a_i'', \ldots \in \mathcal{A}_i$ denote arbitrary distinct elements of $\mathcal{A}_i$, $i = 1,\ldots,n$. Let $\mathcal{A} = \mathcal{A}_1 \times \cdots \times \mathcal{A}_n$ denote the product action space, and let $\mathbf{a} = (a_1,\ldots,a_n)$, $\mathbf{a}' = (a_1',\ldots,a_n')$, $\mathbf{a}'' = (a_1'',\ldots,a_n'')$ denote action profiles in $\mathcal{A}$. For this development it is assumed that each $\mathcal{A}_i$ comprises finitely many elements.

The simplest formalization of preference is with an ordinal binary preference relation that does not account for the intensity, or degree, of preference. The symbol $\succ$ means "is strictly preferred to" and denotes a binary operation that orders elements of some set $\mathcal{A}$ in terms of preference. The expression $a \succ a'$ means that a is superior, by whatever criterion is relevant, to a'. The binary relation $\succ$ is assumed to be *asymmetric*; that is, if $a \succ a'$ then a' is not strictly preferred to a. It is also assumed that $\succ$ is *transitive*: if $a \succ a'$ and $a' \succ a''$, then $a \succ a''$. A relation that is asymmetric and transitive is a *strict partial order*. An ordering relation over $\mathcal{A}$ is *complete* if, for all $a, a' \in \mathcal{A}$ such that $a \neq a'$, either $a \succ a'$ or $a' \succ a$. A relation that is transitive, asymmetric, and complete is called a *strict ordering*.

An associated concept is the indifference relation, denoted $a \sim a'$, meaning that neither action is preferred to the other. Clearly, $\sim$ is *symmetric*, meaning that if $a \sim a'$, then $a' \sim a$. It is often assumed that $\sim$ is transitive, but it is relatively easy to define situations where $a \sim a'$ and $a' \sim a''$ does not imply $a \sim a''$ (see Fishburn (1973) for a discussion). In this treatment, however, $\sim$ is assumed to be transitive.

The strict preference and indifference relations can be combined to form the binary relation $\succsim$, meaning that if $a \succsim a'$, then a is nonstrictly preferred to a'; that is, a is either strictly preferred or indifferent to a'. The binary relation $\succsim$ is *reflexive* if $a \succsim a$ for all $a \in \mathcal{A}$. It is *antisymmetric* if $a \succsim a'$ and $a' \succsim a$ means that $a \sim a'$. It is *transitive* if $a \succsim a'$ and $a' \succsim a''$ means $a \succsim a''$. The relation $\succsim$ is a *partial order* if it is reflexive, antisymmetric, and transitive. It is a *total order* if it is also *complete*, meaning that either $a \succsim a'$ or $a' \succsim a$ (or both) for all $a, a' \in \mathcal{A}$ such that $a \neq a'$. In this treatment, all preference orderings are total orderings unless otherwise stated.

We shall assume, unless otherwise stated, that the binary ordering relations defined so far are with respect to individuals. To indicate the identity of the individual to which the ordering refers, we shall append a subscript to the relation to form the symbol $\succsim_{X_i}$. The domain of an ordering depends on the application. Classical game theory focuses on orderings over elements of the product action space $\mathcal{A} = \mathcal{A}_1 \times \cdots \times \mathcal{A}_n$. Thus, the expression $\mathbf{a} \succsim_{X_i} \mathbf{a}'$ means that X_i nonstrictly prefers $\mathbf{a} = (a_1,\ldots,a_n)$ to $\mathbf{a}' = (a_1',\ldots,a_n')$. The classical social choice context differs from the game-theoretic context in that there is only one action space $\mathcal{A}$ for the group and that only one action can be taken by the group. Under this restriction, the expression $a \succsim_{X_i} a'$ means that X_i nonstrictly prefers a to a'. A social choice problem may be recast in the game format by defining the product action space as the set of all n-tuples of identical elements of the action space, that is, defining $\boldsymbol{\mathcal{A}} = \{(a,\dots,a): a \in \mathcal{A}\}$. The main difference between the game context and the social choice context is that with the former, each X_i is empowered to instantiate its element, a_i, of $\mathbf{a}$, while with the latter, the group as a whole chooses $\mathbf{a} = (a,\dots,a)$.

The multiagent systems context is similar to the game-theoretic context in that each agent possesses its own action space and the benefit to it depends on its ordering defined over the product action space. The main difference between the multiagent systems approach and the game-theoretic approach is that, whereas a classical game does not accommodate a notion of group benefit, an important, and often indispensable, aspect of a multiagent system is that it is designed to accomplish some purpose as a group. In other words, some notion of coordination, as discussed in Section 1.3, is central to its design. Thus, multiagent systems possess characteristics of both game theory and social choice theory: each agent is empowered to instantiate its own decision and the benefit to it depends on the choices of all agents, but some notion of group welfare must also be defined if the system is to be functional in terms of achieving its designed purpose. It is important, indeed, perhaps essential, therefore, that the welfare of the group be defined and reconciled with the welfare of the individuals. This desire, however, is in direct conflict with classical game theory. As pointed out by Shubik (1982): "Game theory makes a special point of *not* requiring 'society' to be a generalized person, capable of making choices and judgments among actions or outcomes on the basis of some sort of welfare function [emphasis in original] " (p. 124).

Shubik is correct in his assertion that decisions must rest with the individuals, and it would be wrong to design a structure such that the group – or, as Raiffa (1968) puts it, the "organization incarnate" – makes the decision. We thus ask the question: Can we formulate an approach that permits a notion of group welfare to emerge while at the same time appropriately honoring a notion of individual welfare? The short answer appears to be no, at least not with the mathematical and logical structures that are used with classical game theory and social choice theory. However, we shall demonstrate how this can be done.

2.1 Complex social models

We begin our discussion by considering the notion of a group preference ordering.

Definition 2.1 Let $\mathcal{X}_n$ be an n-member group, and let the symbol $\succcurlyeq_{\mathcal{X}_n}$ denote a *group-level total preference ordering*. The expression $\mathbf{a} \succcurlyeq_{\mathcal{X}_n} \mathbf{a}'$ means that the group, viewed as a unit and not individually, prefers $\mathbf{a}$ to $\mathbf{a}'$ or is indifferent. $\square$

We use the binary relation symbol $\succcurlyeq_{\mathcal{X}_n}$ rather than $\succsim_{\mathcal{X}_n}$ to emphasize that a group preference ordering is a very different concept than an individual prefer-ence. Classical game theory eschews the notion of group preference ordering, primarily on the grounds that the group is not a sentient individual, or, as Shubik (1982) terms it, a "superplayer," that possesses the power to make and act on decisions. And without a superplayer, it would seem that a group pref-erence ordering, even if it were defined, would be moot. If the group were not empowered to act as if it were a sentient individual, what purpose would such a preference ordering serve?

Our response to this question is that it is not the right one to ask. (However, we shall return to this question in Section 3.4.) A more immediately relevant question is: How should we characterize all of the relationships that arise as social influence propagates through the group? Once social influence has prop-agated through the group, a complex notion of group preference may emerge, but such a notion need not serve directly as the basis for action. Rather, it may serve as a social model that accounts for all of the influence relationships that exist among the players. Once such a model is defined, we may hope to deduce notions of rational behavior for the individuals that take into account the social relationships. Thus, even though the group itself is not empowered to take action as an entity, its interests will be reflected in the decisions made by the individuals.

To gain some insight regarding the structure of such a social model, consider a tennis game where, rather than making shots in ways that make it difficult for the other player to return them, the players make shots in ways that make it easy for the other payer to return them. We may think of each back-and-forth volley as an instantiation of a single-stage game, with the outcome being the shot selection pair at the players' disposal. It is straightforward to infer that a group preference ordering can emerge, even though the group, viewed as an entity, does not dictate the actions. In essence, the group prefers a harmonious result, namely, a successful volley. Although the players may have that objec-tive in mind, the concept of a successful volley is not explicitly encoded in their preference orderings, nor is it under the control of either player – it is an emergent group phenomenon. This notion of group preference has more to do with the compatibility, or suitability, of the players to function appropriately than with the formal aggregation of individual preferences.

In this section we first generalize the notion of individual categorical prefer-ences to include social influence, leading to the formal definition of a conditional preference ordering. We then develop a notion of group preference that is more general than the concept provided by Definition 2.1.

2.1.1 Conjectures

As Arrow's (1951) impossibility theorem attests, reconciling group and individual preference orderings is generally not possible. Arrow's result, however, applies to categorical preferences. Since our objective is to define preference orderings that are not restricted to being categorical, we may entertain the possibility that by relaxing the categorical assumption, we may be able to effect a reconciliation. To do so, we must eschew Friedman's (1961) division of labor hypothesis and move upstream to get closer to the headwaters of the way preferences can be formed. It is reasonable to assume that an individual, acting alone and without any social, economic, or political constraints, would be well justified in defining its preferences solely in its own narrow interest without taking into consideration the effects of its actions on others. However, unless a stakeholder is completely asocial and rejects or lacks the capacity for social interaction, its preferences will be formed by taking into consideration the way it is influenced by the interests, as well as the actions, of others. The question, then, devolves around the issue of how such social influence can be represented and encoded into the preference orderings. To address this question, we introduce the following principle:

Corollary 2.1 Principle 1 (Conditioning) *A stakeholder's preference ordering may be influenced by the preferences of other stakeholders.*

Principle 1 represents an important shift in perspective from classical decision theory. With the classical approach, an individual's preference ordering is with respect to the *instantiation of* actions taken by itself and others. This principle, however, accommodates consideration of the *preferences of others for outcomes* as well as of the actions of others. A convenient way to account for this expanded sphere of interest is to model the preferences of others as the antecedents of hypothetical propositions whose consequents are preference orderings for the individual.

Definition 2.2 A *conjecture* for X_i, denoted $\mathbf{a}_i = (a_{i1}, \ldots, a_{in}) \in \mathcal{A}$, is the antecedent of the hypothetical proposition that X_i considers $\mathbf{a}_i$ as the action profile that will be or should be instantiated.

For any integer m such that $1 \leq m \leq n$, let $\{i_1, \ldots, i_m\}$ denote an increasing m-element subsequence of $\{1, \ldots, n\}$, and let $\mathcal{X}_m = \{X_{i_1}, \ldots, X_{i_m}\}$ be a subgroup of $\mathcal{X}_n$. A *joint conjecture* for $\mathcal{X}_m$, denoted $\boldsymbol{\alpha}_m = \{\mathbf{a}_{i_1}, \cdots, \mathbf{a}_{i_m}\}$, where

$\mathbf{a}_{i_1}, \cdots, \mathbf{a}_{i_m} \in \mathcal{A}$, is a collection of action profiles in $\mathcal{A}^m = \mathcal{A} \times \cdots \times \mathcal{A}$ (m times), where $\mathbf{a}_{i_l}$ is a conjecture for X_{i_l}, $l = 1, \ldots, m$. □

A conjecture could be motivated by the assumed or known social traits of X_{i_l}; it could be externally imposed; it could be motivated by X_{i_l}'s own preferences; or there could be some other reason for forming the hypothesis. Since a conjecture serves only as an antecedent, the motivation behind making it is not relevant. To address relevance, we must examine the consequent.

Definition 2.3 Let $\mathcal{X}_m = \{X_{i_1}, \ldots, X_{i_m}\}$ be a subgroup of $\mathcal{X}_n \backslash \{X_i\}$ that socially influences X_i. A *conditional preference ordering* for X_i given that $\mathcal{X}_m$ jointly conjectures $\alpha_m = \{\mathbf{a}_{i_1}, \cdots, \mathbf{a}_{i_m}\}$, denoted $\succsim_{X_i | \mathcal{X}_m}$, is a binary relation that defines the resulting preference ordering over $\mathcal{A}$ for X_i. The expression $\mathbf{a} | \alpha_m \succsim_{X_i | \mathcal{X}_m} \mathbf{a}' | \alpha_m$ means that X_i (nonstrictly) prefers the profile $\mathbf{a}$ to the profile $\mathbf{a}'$, given that X_{i_l} conjectures $\mathbf{a}_{i_l}$, $l = 1, \ldots, m$. □

Notice that we employ the conditioning symbol "|" that is routinely used in probability theory to indicate conditional probability. In this context, the term on the left-hand side of the conditioning symbol refers to the action profile under consideration by X_i, and the term on the right-hand side refers to the joint conjecture under consideration by $\mathcal{X}_m$. In general, an individual's conditional preference ordering will depend on the subsets of $\mathcal{X}_n$ upon which it conditions. A special case obtains, however, when X_i conditions on the null set of stakeholders.

Definition 2.4 Let $\succsim_{X_i | \varnothing}$, termed a *myopic preference ordering*, denote a preference ordering for X_i that is conditioned on no other stakeholders. We express the condition that $\mathbf{a}$ is myopically preferred (or indifferent) to $\mathbf{a}'$ as $\mathbf{a} | \varnothing \succsim_{X_i | \varnothing} \mathbf{a}' | \varnothing$. □

A myopic preference ordering corresponds to X_i's purely selfish desires under the assumption that it ignores the preferences of all other stakeholders. It is not a categorial preference ordering, per se, however, since it is only one (degenerate) member of a family of conditional preference orderings. For a preference ordering to qualify as categorical, it must be invariant to conditioning on all joint conjectures of all subsets of $\mathcal{X}_n$.

Definition 2.5 If $\succsim_{X_i | \mathcal{X}_m} \equiv \succsim_{X_i | \varnothing}$ for all subsets $\mathcal{X}_m \subset \mathcal{X}_n$, $m < n$ and all $\alpha_m \in \mathcal{A}^m$, then X_i possesses a *categorical preference ordering*. We express the condition that $\mathbf{a}$ is categorically preferred (or indifferent) to $\mathbf{a}'$ as $\mathbf{a} \succsim_{X_i} \mathbf{a}'$. □

Conditional preference orderings provide the mechanism for X_i to extend its sphere of interest beyond itself and to take into consideration the preferences of others when defining its preferences.[1] In particular, it provides a natural way to model altruism. Altruism literally means unselfishness, and it is usually undertaken in the positive sense to mean that one is willing to sacrifice one's own welfare in order to benefit another, but it also applies in the negative sense to a malevolent stakeholder who is willing to sacrifice its welfare in order to injure another. In either case, an altruistic stakeholder, by definition, must take into consideration the preferences of others when defining its own preferences.

By design, categorial preferences accommodate only self-interest, and in that framework, altruism can be accommodated only by redefining self-interest, which is highly problematic, not only philosophically but in practice. It is possible to simulate benevolence or malevolence in particular situations, but such redefinitions tend to be too specific and context dependent for a formal extension of the theory. In particular, it is not possible to distinguish between *categorical altruism*, the willingness always to relinquish one's own self-interest in all situations, and *conditional altruism*, a willingness to relinquish one's own self-interest if and only if (a) the other wishes to take advantage of the offered largess (for benevolent altruism), or (b) the other wishes to act in a way that elicits punishment (malevolent altruism). This more sophisticated expression of altruism is simply not possible with categorical preferences, since a stakeholder's preference ordering is a function of the possible actions of other players, but not their preferences for action. Conditional preferences permit a stakeholder to examine each possible hypothetical situation and adjust its preferences as if the other stakeholders were actually to prefer each of their conjectured outcomes.

Example 2.1 John is taking Mary out to dinner. The choices are Mexican (M), Chinese (C), or Italian (I) fare. In terms of his strictly myopic perspective, John's preference ordering is

$$M \,|\, \varnothing \succ_{John|\varnothing} I \,|\, \varnothing \succ_{John|\varnothing} C \,|\, \varnothing \,.$$

However, John is concerned that Mary enjoys her meal, and thus does not want to abide by his strictly selfish preferences. He thus defines a set of conditional

---

[1] Gintis (2009) offers an alternative model for a player to account for the social propensities of others by introducing a "conjecture probability" that characterizes a player's belief regarding what strategies others will take (his usage of the term "conjecture" is different from our usage). The player then adopts the strategy with the highest expected utility, given the conjecture probability. Gintis's approach, however, retains allegiance to the mathematical structure of categorical utilities and individual rationality, differing mainly through the application of a probability distribution as a means of modulating a player's notion of what is best for itself after taking into consideration what others may prefer to do. The fundamental difference between the two approaches is that, whereas Gintis uses social information to modulate the definition of what is best for a player, given its extant categorical preference ordering, in this book we propose to use social information directly to modulate the player's preference ordering.

preference orderings as

$$M\,|\,M \succ_{John|Mary} I\,|\,M \succ_{John|Mary} C\,|\,M,$$

$$I\,|\,I \succ_{John|Mary} M\,|\,I \succ_{John|Mary} C\,|\,I,$$

$$M\,|\,C \sim_{John|Mary} I\,|\,C \succ_{John|Mary} C\,|\,C\,.$$

Thus, if Mary were to prefer Mexican fare, John's conditional preference ordering would correspond to his myopic preference ordering. If Mary were to prefer Italian fare, John would defer to her preference and reverse his preference regarding Mexican and Italian fare. But there are limits to John's altruism, since he does not care for Chinese food. Consequently, if Mary were to prefer Chinese, he would be indifferent between Mexican and Italian fare, thus expressing a willingness to go with Mary's second choice (either Mexican or Italian), but would prefer either to Chinese.   □

2.1.2 Concordance

Social influence can propagate through a group in complicated ways. X_1 may influence X_2, who may in turn influence X_3, and so on, thereby creating a cascade of social relationships that interconnect the stakeholders in ways that cannot be easily predicted. In this expanded context, it is not sufficient simply to create a payoff array to be subjected to standard solution concepts such as dominance and equilibrium. Instead, we must construct a social model that accounts for all of the interrelationships.

To gain some insight regarding an expanded notion of group preference, let us consider a two-stakeholder group $\{X_1, X_2\}$ with product action space $\mathcal{A} = \{\mathbf{a}, \mathbf{a}', \mathbf{a}''\}$. Suppose the myopic preference orderings are

$$\mathbf{a}\,|\,\varnothing \succ_{X_1|\varnothing} \mathbf{a}'\,|\,\varnothing \succ_{X_1|\varnothing} \mathbf{a}''\,|\,\varnothing,$$

$$\mathbf{a}'\,|\,\varnothing \succ_{X_2|\varnothing} \mathbf{a}\,|\,\varnothing \succ_{X_2|\varnothing} \mathbf{a}''\,|\,\varnothing\,.$$

Now let us consider the following joint conjectures: $\boldsymbol{\alpha}_1 = \{\mathbf{a}, \mathbf{a}'\}$ and $\boldsymbol{\alpha}_2 = \{\mathbf{a}, \mathbf{a}''\}$. With $\boldsymbol{\alpha}_1$, X_1 conjectures what is best for it and next-best for X_2, and X_2 conjectures what is best for it and next-best for X_1. With $\boldsymbol{\alpha}_2$, X_1 conjectures what is best for it and next-best for X_2, but X_2 conjectures what is worst for both. The question is, Which of these two joint conjectures is likely to cause the more severe dispute between the two stakeholders? With $\boldsymbol{\alpha}_1$, although the stakeholders conjecture different outcomes, both of their conjectures avoid the worst outcome. With $\boldsymbol{\alpha}_2$, however, one conjecture at least keeps both from their worst outcome, but the other conjecture consigns both to the worst. It is reasonable to argue that the severity of the dispute is less with $\boldsymbol{\alpha}_1$ than with

α_2. This situation suggests that an ordering can be defined over the set of joint conjectures that characterizes the severity of controversies among the members of the group. Such an ordering represents a measure of *concordance* as the definition of group preference that emerges as the members of the group interact. Concordance, as used in this context, refers to the conformity, suitability, or properness of joint behavior. It has nothing to do with goals, either for the group or for individuals. For example, if X_1 and X_2 were jointly to conjecture $\{a'', a''\}$, there would be no dispute between the two stakeholders, although that joint conjecture would be the worst outcome for both.

For situations where there are common interests among the stakeholders, concordance could be interpreted as harmony, or the similarity of interests. Concordance, however, need not correspond to harmony or cooperation. For example, with an athletic contest, the opposing players do not cooperate in the sense of pursuing a common objective; rather, success in playing the game depends on their opposition to each other. In other words, there is a preference, from the group perspective, for the players to have disputes regarding their desired behavior, and diametrically opposed conjectures would, from a group perspective, have a low degree of controversy. If one player were to "throw" the game by favoring a conjecture similar to the other player's conjecture, the game would be seriously compromised. Thus, even antagonists can behave concordantly.

Definition 2.6 A Let $\mathcal{X}_k = \{X_{i_1}, \ldots, X_{i_k}\}$ be a subgroup of $\mathcal{X}_n$. A *concordant ordering*, denoted $\trianglerighteq_{\mathcal{X}_k}$, is a binary relation that defines a preference ordering of elements of $\mathcal{A}^k$, the set of possible joint conjectures for $\mathcal{X}_k$. The expression $\alpha_k \trianglerighteq \alpha'_k$ means that the joint conjecture α_k is at least as concordant as the joint conjecture α'_k. When $k = 1$, a concordant ordering becomes a conventional categorical ordering $\succsim_{X_k}$ for X_k.   □

Example 2.2 Consider two agents, X_1 and X_2, each of whom wishes to pass through a doorway that is just wide enough to admit two people, as illustrated in Figure 2.1. As they approach the doorway, each has two options: vere to the right (r) or to the left (l). Thus, $\mathcal{A}_1 = \mathcal{A}_2 = \{r, l\}$. Let us consider two games: a cooperative version and a conflictive version:

Cooperative version. The objective of the cooperative version is for both X_1 and X_2 to pass through the doorway. Thus, an appropriate myopic ordering for this version is

$$(r,r)|\varnothing \sim_{X_i|\varnothing} (l,l)|\varnothing \succ_{X_i|\varnothing} (r,l)|\varnothing \sim_{X_i|\varnothing} (l,r)|\varnothing, \ i = 1,2.$$

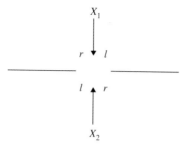

Figure 2.1. The Doorway game.

To illustrate a concordant ordering, let us consider the two joint conjectures $\{(r,r),(r,r)\}$ and $\{(r,l),(l,r)\}$. The joint conjecture $\{(r,r),(r,r)\}$ corresponds to a consensus; there is no controversy. With $\{(r,l),(l,r)\}$, however, there is conflict; hence, the controversy is severe. Thus,

$$\{(r,r),(r,r)\} \rhd_{X_1 X_2} \{(r,l),(l,r)\}$$

would be an appropriate concordant ordering.

Conflictive version. With the conflictive version, the objective is for X_1 and X_2 to block each other's passage; hence, the myopic preference ordering is reversed:

$$(r,l)|\varnothing \sim_{x_i \varnothing} (l,r)|\varnothing \succ_{x_i \varnothing} (r,r)|\varnothing \sim_{x_i \varnothing} (l,l)|\varnothing, \ i = 1,2.$$

Accordingly, the concordant preference ordering is also reversed:

$$\{(r,l),(l,r)\} \rhd_{X_1 X_2} \{(r,r),(r,r)\}.$$

Thus, although the joint conjecture $\{(r,l),(l,r)\}$ represents diametric opposition, from the perspective of the group there is no controversy, while the joint conjecture $\{(r,r),(r,r)\}]\}$ generates a severe controversy. □

A concordance ordering is with respect to the degrees of controversy that arise as the stakeholders each conjecture various outcomes of the game. It provides a complete description of all social relationships that exist within the group, including the propensities to cooperate and compete. it is very different from the group ordering presented in Definition 2.1. In that context, the expression $\mathbf{a} \succcurlyeq_{x_n} \mathbf{a}'$ means that the group, viewed as a unit, prefers the single action profile $\mathbf{a}$ over the action profile $\mathbf{a}'$. A concordant ordering, on the other hand, involves the ordering of sets of profiles, rather than a single profile, and thus cannot be used directly as a basis for group action. Whereas $\succcurlyeq_{x_n}$ could be used to define how the group as a unit can act rationally, the concordant ordering $\rhd_{x_n}$ defines the entire social structure of the group.

If the preference orderings of all players are categorical, the group possesses only minimal sociality, and no explicit social structure exists. Consequently, the concordant preference ordering provides no information that is not already provided by the categorical utilities.

The final step in our development of group-level orderings is to consider the concordant ordering of a subgroup that is influenced by the joint conjecture of another subgroup.

Definition 2.7 A Let $\mathcal{X}_k = \{X_{i_1}, \ldots, X_{i_k}\}$ and $\mathcal{X}_m = \{X_{j_1}, \ldots, X_{j_m}\}$ be two disjoint subgroups of $\mathcal{X}_n$, and let $\boldsymbol{\alpha}_k = \{\mathbf{a}_{i_1}, \ldots, \mathbf{a}_{i_k}\}$ and $\boldsymbol{\alpha}_m = \{\mathbf{a}_{j_1}, \ldots, \mathbf{a}_{j_m}\}$ be joint conjectures for $\mathcal{X}_k$ and $\mathcal{X}_m$, respectively. A *conditional concordant ordering* $\trianglerighteq_{\mathcal{X}_k | \mathcal{X}_m}$ is a binary relation that defines a preference ordering over joint conjectures in $\mathcal{A}^k$ given the joint conjecture $\boldsymbol{\alpha}_m \in \mathcal{A}^m$. The expression $\boldsymbol{\alpha}_k | \boldsymbol{\alpha}_m \trianglerighteq_{\mathcal{X}_k | \mathcal{X}_m} \boldsymbol{\alpha}'_k | \boldsymbol{\alpha}_m$ means that the concordance associated with $\mathcal{X}_k$ jointly conjecturing $\boldsymbol{\alpha}_k$ is at least as great as the concordance for jointly conjecturing $\boldsymbol{\alpha}'_k$, given that $\mathcal{X}_m$ jointly conjectures $\boldsymbol{\alpha}_m$.[2] When $k = 1$, the conditional concordant ordering becomes a conditional preference ordering for X_k, A discussed in Definition 2.3. □

Example 2.3 Consider the group $\{X_1, X_2, X_3\}$, which is tasked to purchase an automobile. They can choose either a foreign-made (F) or a domestic-made (D) vehicle, which will either be a convertible (C), or a sedan (S), and will either be red (R) or green (G). The action sets for $X_i, i = 1, 2, 3$ are $\mathcal{A}_1 = \{F, D\}$, $\mathcal{A}_2 = \{C, S\}$, and $\mathcal{A}_3 = \{R, G\}$. The product action space is $\mathcal{A} = \mathcal{A}_1 \times \mathcal{A}_2 \times \mathcal{A}_3$, that is,

$$\mathcal{A} = \{(F, C, R), (F, C, G), (F, S, R), (F, S, G), (D, C, R), (D, C, G),$$

$$(F, S, R), (D, S, G)\}.$$

The concordant ordering

$$\{(F, C, R), (F, C, R)\} \trianglerighteq_{X_1 X_2} \{(F, C, G), (F, C, G)\}$$

means that if X_1 and X_2 were both to conjecture a foreign-made red convertible, it would be at least as decisive, or compelling, for the subgroup than if they were each were boh to conjecture a foreign-made green convertible.

Now let us consider the expression

$$\{(F, C, R), (D, S, G)\} \triangleright_{X_1 X_2} \{(F, C, G), (D, S, R)\},$$

---

[2] An alternative notation would be to write $\boldsymbol{\alpha}_k \trianglerighteq_{\mathcal{X}_k | \boldsymbol{\alpha}_m} \boldsymbol{\alpha}'_k$ instead of $\boldsymbol{\alpha}_k | \boldsymbol{\alpha}_m \trianglerighteq_{\mathcal{X}_k | \mathcal{X}_m} \boldsymbol{\alpha}'_k | \boldsymbol{\alpha}_m$. However, it seems more clear to include the conditioning quantity as the arguments to the inequality and let the subscript to the ordering symbol denote the ordering and conditioning spaces.

which provides an ordering involving dissimilar conjectures. This ordering relates two hypothesized situations. Let hypothesis H_1 corresponds to the joint conjecture on the left-hand side of this expression: X_1 conjectures a foreign-made red convertible and X_2 conjectures a domestic-made green sedan. Let hypothesis H_2 correspond to the joint conjecture on the right-hand side: X_1 conjectures a foreign-made green convertible, and X_2 conjectures a domestic-made red sedan. Clearly, neither hypothesis corresponds to a consensus for the subgroup; their tastes simply are not compatible. This ordering is a statement that the intensity of incompatibility between X_1 and X_2 under hypothesis H_1 is less than it would be under hypothesis H_2. In other words, if H_2 corresponded to the most-preferred outcomes for X_1 and X_2, the subgroup would be less compatible, or less harmonious, or more conflicted, than if H_1 corresponded to the most-preferred outcomes.

Next, the conditional concordant ordering

$$(F,C,R)|(F,C,R) \trianglerighteq_{X_1|X_2} (D,S,G)|(F,C,R),$$

which means that X_1 prefers a foreign-made red convertible to a domestic-made green sedan, given that X_2 conjectures a foreign-made red convertible. Notice that since the consequent involves only one stakeholder, the conditional concordant ordering becomes a conditional preference ordering, and we may properly replace this expression with

$$(F,C,R)|(F,C,R) \succsim_{X_1|X_2} (D,S,G)|(F,C,R).$$

This example helps to illustrate the role of a conditional ordering. Unlike the conventional categorical preference ordering $\succsim_{X_1}$, which provides the information necessary for stakeholder X_1 to act rationally, the conditional preference ordering $\succsim_{X_1|X_2}$ does not provide sufficient information for X_1 to act rationally. Rather, its role is to characterize the influence relationship that X_2 exerts on X_1. Now suppose that (F,C,R) just happened to be X_2's myopically most-preferred profile. In that case, even if, considered in isolation, X_1 were myopically to prefer a domestic-made green sedan to an foreign-made red convertible, that is, $(D,S,G)|\varnothing \succsim_{X_1|\varnothing} (F,C,R)|\varnothing$, X_1 would be willing to override that preference ordering in deference to the preferences of X_2.

Continuing with this example, the conditional preference ordering

$$(D,S,G)|\{(F,C,R),(D,S,G)\} \succsim_{X_1|X_2X_3} (F,C,R)|\{(F,C,R),(D,S,G)\}$$

means that X_1 prefers a domestic-made green sedan to a foreign-made red convertible, given that X_2 conjectures a foreign-made red convertible and X_3 conjectures a domestic-made green sedan. Thus, the inclusion of X_3's preferences in the conditional preference ordering for X_1 given by this expression reverses the earlier ordering conditioned only on X_2's preferences, an indication that X_1 is more strongly influenced by X_3 than by X_2.

We conclude this example by examining the conditional concordant ordering

$$\{(F,C,R),(F,S,R)\}|(D,S,G) \trianglerighteq_{X_2 X_3 | X_1} \{(D,S,G),(D,C,G)\}|(D,S,G),$$

which means that the subgroup $\{X_2, X_3\}$ is less conflicted, given that X_1 conjectures (D,S,G), for X_2 and X_3 to jointly conjecture $\{(F,C,R),(F,S,R)\}$ than to jointly conjecture $\{(D,S,G),(D,C,G)\}$. $\square$

2.2 Aggregation

Although a concordant ordering may be an appealing concept, its usefulness depends constructing it. Clearly, it cannot be imposed exogenously, since doing so would essentially be no different from imposing a group preference on the players of a classical game. Rather, any notion of group preference must emerge as a consequence of the influence relationships that exist among the stakeholders. This consideration motivates adopting the following principle:

Corollary 2.2 Principle 2 (Endogeny) *A concordant ordering for a group must be determined by the social interactions among its subgroups.*

To motivate our approach for constructing a concordant ordering, we build on existing theory borrowed from an entirely different context. Our point of departure is the observation that using a preference ordering over conjectures as a way of characterizing the benefit of taking action has a natural analogue with using a probabilistic ordering over propositions as a way of characterizing belief regarding which proposition is the true state of nature. To emphasize this distinction, we introduce the term *praxeology*, a neologism created from the Greek roots *praxis* (action) and *logos* (reason). Whereas epistemology involves the classification of propositions on the basis of knowledge and belief, praxeology involves the classification of propositions on the basis of effectiveness and efficiency.[3] In the praxeological context, a conditional ordering is the consequent of a hypothetical proposition whose antecedent is a joint conjecture by other players, and in the epistemological context, a conditional probability is

---

[3] The Oxford English Dictionary defines *praxeology* as "the study of such actions as are necessary in order to give practical effect to a theory or technique; the science of human conduct; the science of efficient action" (Murray et al., 1991).

the consequent of a hypothetical proposition whose antecedent is a claim about the true state of nature for other random phenomena.

We now form an analogy between the epistemological problem of evaluating propositions on the basis of evidence regarding truth and the praxeological issue of evaluating propositions on the basis of usefulness. Let Ω denote a finite set of propositions, exactly one of which is the state of nature; that is, it is true. Let us suppose that $\omega \in \Omega$ is true, and, hence, all $\omega' \in \Omega \setminus \{\omega\}$ are false. This condition induces a preference relation to order elements of Ω, denoted by $\succsim_\Omega^*$, such that $\omega \succ_\Omega^* \omega'$ for all $\omega' \in \Omega \setminus \{\omega\}$ and $\omega' \sim_\Omega^* \omega''$ for all $\omega', \omega'' \in \Omega \setminus \{\omega\}$. We shall term such an ordering a *truth ordering*.

There will often exist conformational evidence to support the assertion that "ω is true." Let $\succsim_\Omega$ denote an ordering of the strength of the confirmation evidence, such that $\omega \succsim_\Omega \omega'$ means that the conformational evidence supporting the statement "ω is true" is at least as strong as the conformational evidence supporting the statement "ω' is true." We will term $\succsim_\Omega$ an *epistemic ordering* relation. Notice, however, that the epistemic ordering relation $\succsim_\Omega$ is very different from the truth ordering relation $\succsim_\Omega^*$. The epistemic ordering is defined by the strength of the conformational evidence ascribed to the propositions, but the truth ordering corresponds to an assumed constraint on the state of nature (that ω is true). From the point of view of an observer, a truth ordering is hypothetical, since the observer generally does not know which of the propositions is, in fact, true. The purpose of such a hypothetical constraint is to enable the observer to evaluate the strength of evidence regarding other phenomena, given that the constraint actually is true.

Let Ω_1 and Ω_2 be two sets of propositions, and let us hypothesize that $\omega_1 \in \Omega_1$ is true, that is, that $\omega_1 \succ_{\Omega_1}^* \omega_1'$ for all $\omega_1' \in \Omega_1$ such that $\omega_1' \neq \omega_1$. Given that hypothesis, we wish to define an epistemic ordering over Ω_2. Let us denote such an ordering by the conditional binary relation symbol $\succsim_{\Omega_2|\Omega_1}$. Essentially, the conditional binary relation is used to construct a hypothetical proposition whose antecedent is the assumption that a given proposition in Ω_1 is true and whose consequent is the resulting ordering over Ω_2.

Example 2.4 Suppose one wishes to determine the weather conditions in a distant city. Let the symbols sy, cy, rt, and dt denote *sunny-yesterday, cloudy-yesterday, rain-today*, and *dry-today*, respectively, and define the spaces $\Omega_1 = \{sy, cy\}$ and $\Omega_2 = \{rt, dt\}$. Suppose one does not have precise knowledge of yesterday's weather state, but is able, perhaps indirectly, to define an epistemic ordering over yesterday's weather states. Let $\succsim_{\Omega_1}$ denote this ordering, and let $\succsim_{\Omega_2|\Omega_1}$ order the degree of belief over Ω_2 regarding today's weather state given yesterday's hypothesized true weather state. Thus, the expression $dt|sy \succsim_{\Omega_2|\Omega_1}$

$rt|sy$ means, given that the state *sunny-yesterday* is true, that the state *dry-today* is at least as likely as the state *rain-today*.

The ordering relations $\succsim_{\Omega_1}$ and $\succsim_{\Omega_2|\Omega_1}$ are functionally different. Whereas the ordering $\succsim_{\Omega_1}$ provides an indication of relative belief and may, therefore, be used as the basis of a rational decision regarding the state of yesterday's weather, the ordering $\succsim_{\Omega_2|\Omega_1}$ provides a measure of conditional relative belief and does not, by itself, provide a basis for a rational decision regarding the state of today's weather. However, by using the two orderings together, it may be possible (as we shall eventually see) to make a rational prediction regarding today's weather.   □

In general, for any two sets of epistemological propositions Ω_1 and Ω_2, we may define binary relations $\succsim_{\Omega_i}$, $i = 1, 2$ to express ordinal relationships concerning the strength of confirmational evidence regarding the truth status of the elements of Ω_i. Also, we may define conditional binary relations $\succsim_{\Omega_i|\Omega_j}$, $i, j = 1, 2$, $j \neq i$, to order the strength of confirmational evidence regarding the truth status of elements in Ω_i given the hypothesis that $\omega_j \in \Omega_j$ is true. Thus, we may construct general hypothetical statements of the form $\omega_i|\omega_j \succsim_{\Omega_i|\Omega_j} \omega_i'|\omega_j$ for $j \neq i$ to mean that the confirmational evidence that ω_i is the state of nature in Ω_i is at least as strong as is the confirmational evidence that ω_i' is the state of nature in Ω_i, given that ω_j is the state of nature in Ω_j.

Let us now move to a praxeological context, and let $\mathcal{A}$ denote a finite set of alternatives for a stakeholder X. In this context the notion that there is a "true" alternative does not apply. However, just as the observer in an epistemological context may invoke a hypothetical truth constraint in order to induce a truth ordering, the stakeholder in a praxeological context may invoke a hypothetical constraint (an antecedent) that any given alternative is the one that should be implemented – a *conjecture*. A conjecture a induces a preference ordering over the elements of $\mathcal{A}$, denoted $\succsim_X^*$, such that $a \succ_X^* a'$ for all $a' \in \mathcal{A}\backslash\{a\}$ and $a' \sim_X^* a''$ for all $a', a'' \in \mathcal{A}\backslash\{a\}$. We shall term such an ordering a *conjecture ordering*, as a parallel to the truth ordering in the epistemological context.

Now suppose information exists concerning the benefit to the stakeholder regarding the alternatives. This information will induce an ordering $\succsim_X$ over the alternatives such that $a \succsim_X a'$ means that the benefit of instantiating a is at least as great as the benefit of instantiating a'. We shall term $\succsim_X$ a *praxeic ordering* relation. However, the praxeic ordering relation $\succsim_X$ is very different from the conjecture ordering relation $\succsim_X^*$. The praxeic ordering is defined by the strengths of the valuations ascribed to the alternatives, and the conjecture ordering corresponds to a constraint defined by the stakeholder. Such a constraint is hypothetical, since the stakeholder is not required actually to abide by that constraint. The purpose of such a constraint is to enable other stakeholders to characterize the strength of the evaluations regarding the alternatives, given that the hypothetical constraint is instantiated.

Just as, in a epistemological context where we may condition the strength of confirmational evidence on a truth assertion, in a praxeological context we may condition the strength of evidence on a conjecture. Let $\mathcal{A}_1$ and $\mathcal{A}_2$ be action sets for stakeholders X_1 and X_2. We may then form a hypothetical proposition whose antecedent is a conjecture $a_1 \in \mathcal{A}_1$ and whose consequent is a conditional praxeic ordering over $\mathcal{A}_2$ of the form $a_2|a_1 \succsim_{X_2|X_1} a'_2|a_1$, which means that the benefit to X_2 of instantiating a_2 is at least as great as the benefit of instantiating a'_2, given that a_1 is a conjecture in $\mathcal{A}_1$.

The analogy between the epistemological and praxeological suggests that the way of thinking regarding epistemological situations may also apply to praxeological situations. While this analogy is not a proof, it does serve to motivate a deeper investigation, to which we now turn.

2.2.1 Ordinal aggregation

To aggregate is to combine the preference orderings of individuals or small groups to form an ordering for a larger group. In the social choice and multi-agent contexts, some form of aggregation is relevant, since a group-level choice is desired. But if we insist on the classical model of categorical preferences, then it is difficult to see how a meaningful notion of group preference could be well defined, even though a group-level choice may be rendered. In such an environment, each participant is concerned with, and only with, its own interests. Thus, any concept of group preference would be the result of nothing more than a coincidence of compatible preferences. However, if the stringent requirement of categorical preferences were relaxed, then it may be more than merely coincidence that a notion of group-level desires could emerge from the aggregation of individual preferences. Such aggregation, however, must not be exogenously imposed by some entity outside the society. In order to conform to the endogeny principle, if a group preference ordering is to exist, it must emerge from the way the individuals interact.

To consider ways to define endogenous aggregation, we may gain some insight by examining how aggregation takes place in an epistemological setting, where it is often necessary to combine information from different sources to form a complete assessment of the state of nature. Let us reconsider the weather forecasting scenario defined in Example 2.4. Suppose we wish to determine an epistemic ordering for the state of the weather in a distant city today, given an epistemic ordering for weather yesterday and the degree of belief regarding weather conditions today given conditions yesterday. One way to proceed is to combine this information in two steps. The first step is to expand the scope of interest to consider the weather conditions on both days simultaneously and determine an epistemic ordering over all joint conditions (e.g., cloudiness yesterday and rain today). Let us call this the *aggregation step*. The second step is to condense this information to form an epistemic ordering over the state of the weather today. Let us call this second step the *marginalization step*.

We first concentrate on the aggregation phase of this enterprise. Suppose we are given two sets of propositions, Ω_1 and Ω_2, and we are also given a binary relation $\succsim_{\Omega_1}$ and a conditional binary relation $\succsim_{\Omega_2|\Omega_1}$. Can these orderings be combined to form an ordering for the ordered pair of propositions $(\omega_1, \omega_2) \in \Omega_1 \times \Omega_2$? To do so, there must exist an ordinal epistemic aggregation function f of these orderings of the form

$$\succsim_{\Omega_1\Omega_2} = f\left(\succsim_{\Omega_1}, \succsim_{\Omega_2|\Omega_1}\right). \tag{2.1}$$

Is such a function well defined? To address this question, we consider the argument put forth by Jaynes (2003). For the joint proposition $(\omega_1, \omega_2) \in \Omega_1 \times \Omega_2$ to be true, it must hold that $\omega_1 \in \Omega_1$ is true. But if ω_1 is true, then for (ω_1, ω_2) to be true, it must hold that ω_2 is true given that ω_1 is true. However, if ω_1 is false, then (ω_1, ω_2) is false, regardless of the truth status of ω_2. If we first consider the truth of ω_1, then the epistemic ordering of ω_2 will be relevant to the epistemic ordering of (ω_1, ω_2) only if ω_1 is true. Thus, knowledge of the epistemic ordering $\succsim_{\Omega_1}$ over Ω_1 and the family of conditional epistemic orderings $\succsim_{\Omega_2|\Omega_1}$ over Ω_2 for each $\omega_1 \in \Omega_1$ is sufficient to determine the epistemic ordering $\succsim_{\Omega_1\Omega_2}$ over $\{\Omega_1, \Omega_2\}$, and the ordering $\succsim_{\Omega_2}$ is not needed.

To explore the possibility of identifying a function f as defined in (2.1), we need to define the ordering between any two pairs (ω_1, ω_2) and (ω_1', ω_2'). Suppose that $\omega_2|\omega_1 \succsim_{\Omega_2|\Omega_1} \omega_2'|\omega_1$. If $\omega_1 \succsim_{\Omega_1} \omega_1'$, then it is reasonable that $(\omega_1, \omega_2) \succsim_{\Omega_1\Omega_2} (\omega_1', \omega_2')$. However, if $\omega_1' \succsim_{\Omega_1} \omega_1$, then the ordering of (ω_1, ω_2) relative to (ω_1', ω_2') is not clear. Thus, as long as we deal only with ordinal values, we can make no more progress in defining the function f. To proceed further, we must add more structure to the orderings. Specifically, we must account for the strength of preference. In other words, we will need to define cardinal orderings over Ω_1 and a family of conditional cardinal orderings over Ω_2 for each $\omega_1 \in \Omega_1$. Before taking this step, however, let us first discuss ordinal aggregation in the praxeological context.

The argument regarding the aggregation of cardinal and conditional preferences as defined earlier in the epistemological context may also be applied to the praxeological context as follows. Let $\mathcal{X}_m$ and $\mathcal{X}_k$ be arbitrary disjoint subgroups of $\mathcal{X}_n$. For $\{\boldsymbol{\alpha}_m, \boldsymbol{\alpha}_k\}$ to be a joint conjecture in $\mathcal{A}^{mk} = \mathcal{A}^m \times \mathcal{A}^k$, it is necessary that $\boldsymbol{\alpha}_k$ be a joint conjecture in $\mathcal{A}^k$. However, if $\boldsymbol{\alpha}_k$ is not a joint conjecture, then $\{\boldsymbol{\alpha}_m, \boldsymbol{\alpha}_k\}$ is not a joint conjecture, regardless of whether or not $\boldsymbol{\alpha}_m$ is a joint conjecture. Thus, when considering the concordant ordering of $\{\boldsymbol{\alpha}_m, \boldsymbol{\alpha}_k\}$, if a joint conjecture $\boldsymbol{\alpha}_k$ is considered first, then the concordant ordering over $\mathcal{A}^m$ will be relevant only if $\boldsymbol{\alpha}_k$ is a joint conjecture. Consequently, the concordant ordering $\unrhd_{\mathcal{X}_k}$ over $\mathcal{A}^k$ and the family of conditional concordant orderings $\unrhd_{\mathcal{X}_m|\mathcal{X}_k}$ over $\mathcal{A}^m$ for each $\boldsymbol{\alpha}_k \in \mathcal{A}^k$ is sufficient to define the concordant ordering $\unrhd_{\mathcal{X}_m \mathcal{X}_k}$ over $\mathcal{A}^{mk}$. Thus, analogous to the way epistemic orderings may be aggregated·in the epistemological domain according to (2.1), the concordant orderings may be aggregated if there exists a praxeological aggregation function F such that

$$\unrhd_{\mathcal{X}_m \mathcal{X}_k} = F(\unrhd_{\mathcal{X}_k}, \unrhd_{\mathcal{X}_m|\mathcal{X}_k}), \tag{2.2}$$

where $\unrhd_{\mathcal{X}_m \mathcal{X}_k}$ is a conjecture ordering over $\mathcal{A}^{mk}$. However, as within the epistemological context, it is impossible to define such a function with only ordinal preference relations. Consequently, we must turn our attention to cardinal orderings.

2.2.2 Cardinal aggregation

Epistemological cardinal aggregation

In the epistemological context, where the goal is to combine confirmational evidence to support a belief that a given state of nature is true, it is natural to ascribe numerical values to the evidence as a measure of its strength. Although there is not a unique way to ascribe value, nor is there a unique way to combine evidence, three approaches have gained traction as reasonable ways to aggregate evidence. They are (a) probability theory, (b) possibility theory (Zadeh, 1978), and (c) Dempster-Shafer theory (Shafer, 1976). Probability theory has long been established as an important and useful mathematical model of uncertainty. Possibility theory and Dempster-Shafer theory are relatively recent approaches that have been advanced as alternatives to probability theory to deal with situations where there is insufficient information to define a probabilistic model. Although these latter two models are of considerable interest, in this book we focus on probability theory as a means of mathematically characterizing confirmation evidence regarding the truth of propositions.

Let us now return to the problem of finding the function f described in (2.1). Given the orderings $\succsim_{\Omega_1}$ and $\succsim_{\Omega_2|\Omega_1}$, suppose there exists a probability mass function p_{Ω_1} such that $p_{\Omega_1}(\omega_1) > p_{\Omega_1}(\omega_1')$ if and only if $\omega_1 \succ_{\Omega_1} \omega_1'$ and $p_{\Omega_1}(\omega_1) = p_{\Omega_1}(\omega_1')$ if and only if $\omega_1 \sim_{\Omega_1} \omega_1'$. Also, suppose there exists a conditional probability mass function $p_{\Omega_2|\Omega_1}$ such that $p_{\Omega_2|\Omega_1}(\omega_2|\omega_1) > p_{\Omega_2|\Omega_1}(\omega_2'|\omega_1)$

if and only if $\omega_2|\omega_1 \succ_{\Omega_2|\Omega_1} \omega_2'|\omega_1$ and $p_{\Omega_2|\Omega_1}(\omega_2|\omega_1) = p_{\Omega_2|\Omega_1}(\omega_2'|\omega_1)$ if and only if $\omega_2|\omega_1 \sim_{\Omega_2|\Omega_1} \omega_2'|\omega_1$. Given the probability structure, we immediately see that the aggregation function is defined by the chain rule of probability

$$p_{\Omega_1\Omega_2}(\omega_1,\omega_2) = f\big(p_{\Omega_1}(\omega_1), p_{\Omega_2|\Omega_1}(\omega_2|\omega_1)\big) = p_{\Omega_2|\Omega_1}(\omega_2|\omega_1)p_{\Omega_1}(\omega_1), \quad (2.3)$$

which defines the ordering $\succsim_{\Omega_1\Omega_2}$ such that $(\omega_1,\omega_2) \succ_{\Omega_1\Omega_2} (\omega_1',\omega_2')$ if and only if $p_{\Omega_1\Omega_2}(\omega_1,\omega_2) > p_{\Omega_1\Omega_2}(\omega_1',\omega_2')$ and $(\omega_1,\omega_2) \sim_{\Omega_1\Omega_2} (\omega_1',\omega_2')$ if and only if $p_{\Omega_1\Omega_2}(\omega_1,\omega_2) = p_{\Omega_1\Omega_2}(\omega_1',\omega_2')$.

The joint probability mass function defined by (2.3) provides the aggregation step of combining information from different sources. Thus, returning to the weather prediction problem discussed in Example 2.4, the joint probability of $(\omega_1,\omega_2) \in \{sy,cy\} \times \{rt,dt\}$ is given by

$$p_{\Omega_1\Omega_2}(\omega_1,\omega_2) = p_{\Omega_2|\Omega_1}(\omega_2|\omega_1)p_{\Omega_1}(\omega_1).$$

Once the aggregation step is completed, we may apply the marginalization step to compute the probability of the current weather conditions as the marginal probability mass function

$$p_{\Omega_2}(\omega_2) = \sum_{\omega_1} p_{\Omega_1\Omega_2}(\omega_1,\omega_2),$$

which defines an epistemic ordering $\succsim_{\Omega_2}$ over today's weather states.

Praxeological cardinal aggregation
There are good reasons why the theorems should all be easy and the definitions hard.
— Michael Spivak
Calculus on Manifolds (W. A. Benjamin, 1965)

We now introduce a number of concepts that lead to our central result, the aggregation theorem, which establishes the methodology for the aggregation of individual conditional preference orderings to lead, ultimately, to the notion of an emergent group-level preference ordering and to individual preference orderings. To proceed, however, we must represent preferences numerically, rather than with binary preference relations. Such numerical representations are termed *utilities*.[4] In general, utilities may be ordinal or cardinal. For example,

---

[4] The development of utility theory presented here is under the condition of deterministic decision making. It should be mentioned that an axiomatic treatment of utility theory, as discussed in Luce and Raiffa (1957), involves decision making under risk, where utilities are defined over lotteries rather than over deterministic alternatives. Some authors (see Keeney and Raiffa, 1993) use the terminology value function or preference function for the deterministic case, whereas others use the term *utility* for both the deterministic and random cases, as we shall see.

suppose $\mathbf{a} \succ_{X_i} \mathbf{a}'$. One might arbitrarily associate the integers 2 and 1 with $\mathbf{a}$ and $\mathbf{a}'$, respectively, simply to indicate that $\mathbf{a}$ is more preferred than $\mathbf{a}'$, but that does not necessarily imply that $\mathbf{a}$ is preferred twice as strongly as $\mathbf{a}'$. Thus, the mere application of numerical values does not mean that they designate the degree of preference. Such a utility would be ordinal. We shall insist, however, that all utilities under consideration must be cardinal; that is, the utilities provide a measure of the degree, or intensity, of preference.

Although cardinal utilities are used to define the degree of preference, the units in which the preferences are expressed are arbitrary. Accordingly, utilities may be scaled and biased by arbitrary positive affine transformations without loss of generality. Thus, if u is a utility over $\mathcal{A}$, then the transformation $u'(\mathbf{a}) = cu(\mathbf{a}) + b$, where $c > 0$ and b is arbitrary, is also a utility, simply expressed in different units.

Since the solution concepts of classical game theory are defined in terms of individual rationality, all of the utilities in that context are restricted to characterizing preferences for individuals. With our treatment, however, we intend to define preference orderings for both individuals and groups, and thus we need to introduce the concepts of conditional utility, concordant utility, and conditional concordant utility.

Definition 2.8 Let $\mathcal{X}_m = \{X_{j_1}, \ldots, X_{j_m}\}$ be a subgroup of $\mathcal{X}_n$ and let $\boldsymbol{\alpha}_m = \{\mathbf{a}_{j_1}, \cdots, \mathbf{a}_{j_m}\}$ be a joint conjecture for $\mathcal{X}_m$. For $X_i \notin \mathcal{X}_m$, suppose there exists a conditional ordering $\succsim_{X_i | \mathcal{X}_m}$ over $\mathcal{A}$ for each $\boldsymbol{\alpha}_m \in \mathcal{A}^m$ (see Definition 2.3). A *conditional utility*[5] $u_{X_i | \mathcal{X}_m}(\cdot | \boldsymbol{\alpha}_m)$ is a real-valued function defined over $\mathcal{A}$ such that $u_{X_i | \mathcal{X}_m}(\mathbf{a} | \boldsymbol{\alpha}_m) > u_{X_i | \mathcal{X}_m}(\mathbf{a}' | \boldsymbol{\alpha}_m)$ if and only if $\mathbf{a} \succ_{X_i | \mathcal{X}_m} \mathbf{a}'$ and $u_{X_i | \mathcal{X}_m}(\mathbf{a} | \boldsymbol{\alpha}_m) = u_{X_i | \mathcal{X}_m}(\mathbf{a}' | \boldsymbol{\alpha}_m)$ if and only if $\mathbf{a} \sim_{X_i | \mathcal{X}_m} \mathbf{a}'$. If these conditions hold, we say that $u_{X_i | \mathcal{X}_m}$ *agrees* with $\succsim_{X_i | \mathcal{X}_m}$. Without loss of generality, we may assume that conditional utilities are normalized; that is,

$$u_{X_i | \mathcal{X}_m}(\mathbf{a} | \boldsymbol{\alpha}_m) \geq 0 \ \forall \mathbf{a} \in \mathcal{A}, \ \forall \boldsymbol{\alpha}_m \in \mathcal{A}^m$$

and

$$\sum_{\mathbf{a}} u_{X_i | \mathcal{X}_m}(\mathbf{a} | \boldsymbol{\alpha}_m) = 1 \ \forall \boldsymbol{\alpha}_m \in \mathcal{A}^m .$$

---

[5] A concept of a conditional utility that is syntactically similar to our approach is the notion of attribute dominance introduced by Abbas and Howard (2005) and Abbas (2009), who construct a conditional utility function as the ratio of a bivariate utility and a marginal utility, where the marginal corresponding to an individual attribute is the bivariate utility evaluated at the maximally preferred consequence of the other attribute. This approach requires the evaluation of the simultaneous instantiation of the consequences corresponding to both attributes, whereas the conditional utilities defined herein provide an evaluation of the instantiation of the individual actions conditioned on the hypothetical instantiation of other actions. The two concepts have different definitions and different interpretations.

To make this notation explicit, $u_{X_i|\mathcal{X}_m}(\mathbf{a}|\boldsymbol{\alpha}_m) > u_{X_i|\mathcal{X}_m}(\mathbf{a}'|\boldsymbol{\alpha}_m)$ if and only if $\mathbf{a}|\boldsymbol{\alpha}_m \succ_{X_i|\mathcal{X}_m} \mathbf{a}'|\boldsymbol{\alpha}_m$ and $u_{X_i|\mathcal{X}_m}(\mathbf{a}|\boldsymbol{\alpha}_m) = u_{X_i|\mathcal{X}_m}(\mathbf{a}'|\boldsymbol{\alpha}_m)$ if and only if $\mathbf{a}|\boldsymbol{\alpha}_m \sim_{X_i|\mathcal{X}_m} \mathbf{a}'|\boldsymbol{\alpha}_m$. □

As discussed in Section 2.1.1, we must consider the case where X_i conditions on the null set of stakeholders.

Definition 2.9 Let $\succsim_{X_i|\varnothing}$ be a myopic preference ordering for X_i. A *myopic utility* for X_i is a function $u_{X_i|\varnothing}(\cdot|\varnothing)$ that agrees with $\succsim_{X_i|\varnothing}$. Thus, $\mathbf{a}$ is myopically (nonstrictly) preferred to $\mathbf{a}'$ by X_i if $u_{X_i|\varnothing}(\mathbf{a}|\varnothing) \geq u_{X_i|\varnothing}(\mathbf{a}'|\varnothing)$. □

Definition 2.10 Suppose X_i possesses a categorical preference ordering $\succsim_{X_i}$, that is, its conditional preference ordering is invariant. A *categorical utility* is a real-valued function u_{X_i} defined over $\mathcal{A}$ that agrees with $\succsim_{X_i}$. Thus, $\mathbf{a}$ is categorically (nonstrictly) preferred to $\mathbf{a}'$ by X_i if $u_{X_i}(\mathbf{a}) \geq u_{X_i}(\mathbf{a}')$. □

Definition 2.11 Let $\mathcal{X}_k = \{X_{i_1}, \ldots, X_{i_k}\}$ be a subgroup of $\mathcal{X}_n$ and let $\unrhd_{\mathcal{X}_k}$ be a concordant ordering over $\mathcal{A}^k$ (see Definition 2.6). A *concordant utility* $U_{\mathcal{X}_k}$ is a real-valued function defined over $\mathcal{A}^k$ that agrees with $\unrhd_{\mathcal{X}_k}$. When $k = 1$, the concordant utility becomes the conventional cardinal utility u_{X_k} for X_k. Without loss of generality, we may assume that concordant utilities are normalized; that is,

$$U_{\mathcal{X}_k}(\boldsymbol{\alpha}_k) \geq 0 \ \forall \boldsymbol{\alpha}_k \in \mathcal{A}^k$$

and

$$\sum_{\boldsymbol{\alpha}_k} U_{\mathcal{X}_k}(\boldsymbol{\alpha}_k) = 1.$$

For convenience, we shall use the notation $U_{\mathcal{X}_n}$ and $U_{X_1 \cdots X_n}$ interchangeably. □

To help understand the role of the concordant utility, let us examine some specific cases. First, suppose $U_{\mathcal{X}_k}(\boldsymbol{\alpha}_k) = 0$. For this joint conjecture, the level of controversy is maximum; there is not the slightest degree of concordance among the participants. Next, suppose $U_{\mathcal{X}_k}(\boldsymbol{\alpha}_k) = 1$ and thus $U_{\mathcal{X}_k}(\boldsymbol{\alpha}_k') = 0$ for all $\boldsymbol{\alpha}_k' \neq \boldsymbol{\alpha}_k$. For this case, maximum concordance occurs when each player conjectures its component of $\boldsymbol{\alpha}_k$. Such a condition would not necessarily mean that all of the components of $\boldsymbol{\alpha}_k$ are the same, although such a situation would indicate that an overwhelming consensus would exist. It could conceivably occur, however, that maximum concordance for the group could obtain when the players's individual interests are diametrically opposed, such as might happen within an athletic context. As a final extreme case, suppose $U_{\mathcal{X}_k}(\boldsymbol{\alpha}_k) = U_{\mathcal{X}_k}(\boldsymbol{\alpha}'_k)$ for all $\boldsymbol{\alpha}'_k$, which would signify complete indifference for the group, although the individuals may have nonuniform preferences.

Definition 2.12 Let $\mathcal{X}_k = \{X_{i_1}, \ldots, X_{i_k}\}$ and $\mathcal{X}_m = \{X_{j_1}, \ldots, X_{j_m}\}$ be two disjoint subgroups of $\mathcal{X}_n$, and let $\boldsymbol{\alpha}_k = \{\mathbf{a}_{i_1}, \ldots, \mathbf{a}_{i_k}\}$ and $\boldsymbol{\alpha}_m = \{\mathbf{a}_{j_1}, \ldots, \mathbf{a}_{j_m}\}$ be joint conjectures for $\mathcal{X}_k$ and $\mathcal{X}_m$, respectively. Let $\trianglerighteq_{\mathcal{X}_k | \mathcal{X}_m}$ be a conditional concordant ordering over $\mathcal{A}^k$ (see Definition 2.7). For each $\boldsymbol{\alpha}_m \in \mathcal{A}^m$, a *conditional concordant utility* is a real-valued function $U_{\mathcal{X}_k | \mathcal{X}_m}(\cdot | \boldsymbol{\alpha}_m)$ defined over $\mathcal{A}^k$ that agrees with $\trianglerighteq_{\mathcal{X}_k | \mathcal{X}_m}$ for each $\boldsymbol{\alpha}_m$. When $k = 1$, the conditional concordant utility becomes a conditional utility for X_k as defined in Definition 2.8. Without loss of generality, we may assume that conditional concordant utilities are normalized; that is,

$$U_{\mathcal{X}_k | \mathcal{X}_m}(\boldsymbol{\alpha}_k | \boldsymbol{\alpha}_m) \geq 0 \ \forall \boldsymbol{\alpha}_k \in \mathcal{A}^k, \ \forall \boldsymbol{\alpha}_m \in \mathcal{A}^m$$

and

$$\sum_{\boldsymbol{\alpha}_k} U_{\mathcal{X}_k | \mathcal{X}_m}(\boldsymbol{\alpha}_k | \boldsymbol{\alpha}_m) = 1 \ \forall \boldsymbol{\alpha}_m \in \mathcal{A}^m.$$

To make the notation specific, $U_{\mathcal{X}_k | \mathcal{X}_m}(\boldsymbol{\alpha}_k | \boldsymbol{\alpha}_m) > U_{\mathcal{X}_k | \mathcal{X}_m}(\boldsymbol{\alpha}'_k | \boldsymbol{\alpha}_m)$ if and only if $\boldsymbol{\alpha}_k | \boldsymbol{\alpha}_m \vartriangleright_{\mathcal{X}_k | \mathcal{X}_m} \boldsymbol{\alpha}'_k | \boldsymbol{\alpha}_m$, and $U_{\mathcal{X}_k | \mathcal{X}_m}(\boldsymbol{\alpha}_k | \boldsymbol{\alpha}_m) = U_{\mathcal{X}_k | \mathcal{X}_m}(\boldsymbol{\alpha}'_k | \boldsymbol{\alpha}_m)$ if and only if $\boldsymbol{\alpha}_k | \boldsymbol{\alpha}_m \sim_{\mathcal{X}_k | \mathcal{X}_m} \boldsymbol{\alpha}'_k | \boldsymbol{\alpha}_m$. $\square$

Notice that we may view the categorical utility u_{X_i} as a concordant utility U_{X_i} defined over the singleton subgroup $\{X_i\}$, and we may view the conditional utility $u_{X_i | \mathcal{X}_k}$ as a conditional concordant utility $U_{X_i | \mathcal{X}_k}$ defined over the singleton subgroup $\{X_i\}$ conditioned on the subgroup $\mathcal{X}_k$.

2.2.3 Synthesis

Our task now turns to the issue of how to synthesize a concordant utility for a group, given concordant utilities of its subgroups. Specifically, given two disjoint subgroups $\mathcal{X}_k$ and $\mathcal{X}_m$, where $1 \leq k, m < n$, we wish to synthesize the concordant utility of $\mathcal{X}_k \cup \mathcal{X}_m$. Motivated by a desire to formulate an approach that represents a significant generalization that has practical application and yet is sufficiently simple to permit a tractable solution, we propose the following principle:

Corollary 2.3 Principle 3 (Acyclicity) *No cycles may occur in the social influence relationships among the stakeholders.*

Acyclicity essentially means that given two disjoint subgroups $\mathcal{X}_k$ and $\mathcal{X}_m$ of $\mathcal{X}_n$, it cannot happen that, simultaneously, $\mathcal{X}_m$ directly influences $\mathcal{X}_k$ and, simultaneously, $\mathcal{X}_k$ directly influences $\mathcal{X}_m$. The fact that cycles are not permitted does reduce the generality of the model. (In Chapter 6, however, we present a structure that partially overcomes this limitation.) Nevertheless, restricting to one-way direct social influence relationships is a significant generalization of the classical approaches, which assume that all utilities are categorical and, hence, explicit social influence relationships do not even exist.

Many social organizations possess acyclical structures. Business firms and military organizations possess hierarchical structures, with CEOs and commanding officers at the top with tree structures of subordinates beneath them. Accordingly, social influence in such groups is modeled as a top-down structure.

The most fundamental consequence of acyclicity is that there must exist at least one stakeholder who possesses a categorical preference ordering. In Section 2.4 we will also see that acyclicity leads to a powerful and intuitively pleasing graphical representation of multistakeholder groups.

Let $\mathcal{X}_k$ and $\mathcal{X}_m$ denote two disjoint subgroups of $\mathcal{X}_n$, with concordant utilities $U_{\mathcal{X}_k}$ and $U_{\mathcal{X}_m|\mathcal{X}_k}$. Let $U_{\mathcal{X}_k \mathcal{X}_m}$ denote the concordant utility of $\mathcal{X}_k \cup \mathcal{X}_m$. Our goal is to define an *aggregation function* F such that

$$U_{\mathcal{X}_k \mathcal{X}_m}(\boldsymbol{\alpha}_k, \boldsymbol{\alpha}_m) = F[U_{\mathcal{X}_k}(\boldsymbol{\alpha}_k), U_{\mathcal{X}_m|\mathcal{X}_k}(\boldsymbol{\alpha}_m | \boldsymbol{\alpha}_k)]. \qquad (2.4)$$

Drawing on the analogy between epistemological and praxeological orderings, it is temping simply to appropriate the probabilistic structure and apply the chain rule without further justification. Such an action would be premature, however, since there are significant differences between the two domains that must be addressed before such a move can be justified. An important issue in this regard is that, when social relationships exist among the members of a group, there generally will not be a unique way to represent them mathematically. To illustrate, let $\{X_1, X_2\}$ be a two-member group such that X_1 possesses a categorical utility u_{X_1} and X_2 possesses a conditional utility $u_{X_2|X_1}$. Let us consider the aggregation of the utilities u_{X_1} and $u_{X_2|X_1}$ to form a concordant utility for $\{X_1, X_2\}$. As thus framed, the goal is to define F such that

$$U_{X_1 X_2}(\mathbf{a}_1, \mathbf{a}_2) = F[u_{X_1}(\mathbf{a}_1), u_{X_2|X_1}(\mathbf{a}_2 | \mathbf{a}_1)].$$

Notice that the result of this aggregation is a concordant utility, since it involves the joint conjecture $\{\mathbf{a}_1, \mathbf{a}_2\}$.

Now let us suppose that there is a well-defined social relationship between X_1 and X_2 such that when defining their preferences, they both take into consideration that, ultimately, they will be operating in a group environment and not in isolation. Specifically, we assume that (a) neither of these participants is so obdurate and inflexible that it will not take into consideration, in any degree whatsoever, the interests of others when defining its own preferences and (b) exactly the same information is available to both players as with the original framing. Under these conditions, it is possible to reframe the scenario by X_2 defining a categorical utility u_{X_2} and X_1 defining a conditional utility $u_{X_1|X_2}$. Under this alternate framing, the aggregation problem requires

$$U_{X_2 X_1}(\mathbf{a}_2, \mathbf{a}_1) = F[u_{X_2}(\mathbf{a}_2), u_{X_1|X_2}(\mathbf{a}_1|\mathbf{a}_2)].$$

A highly desirable condition is that $U_{X_1 X_2}(\mathbf{a}_1, \mathbf{a}_2) = U_{X_2 X_1}(\mathbf{a}_2, \mathbf{a}_1)$ or, equivalently, that

$$F[u_{X_1}(\mathbf{a}_1), u_{X_2|X_1}(\mathbf{a}_2|\mathbf{a}_1)] = F[u_{X_2}(\mathbf{a}_2), u_{X_1|X_2}(\mathbf{a}_1|\mathbf{a}_2)].$$

This desire motivates the following principle:

Corollary 2.4 Principle 4 (Exchangeability) *If a multistakeholder decision problem can be framed in more than one way using exactly the same information, all such framings should yield the same concordant preference ordering.*

The essence of this principle is that the social model of a group should be invariant to the way it is synthesized, assuming that all of the relevant information is preserved, albeit encoded in different ways.

Definition 2.13   Let $\mathcal{X}_k$ and $\mathcal{X}_m$ be two disjoint subgroups of $\mathcal{X}_n$, and suppose there exist two framings of the preferences and relationships between the two subgroups of the forms $\{U_{\mathcal{X}_k}, U_{\mathcal{X}_m|\mathcal{X}_k}\}$ and $\{U_{\mathcal{X}_m}, U_{\mathcal{X}_k|\mathcal{X}_m}\}$. Let F be an aggregation function such that

$$U_{\mathcal{X}_k \mathcal{X}_m}(\boldsymbol{\alpha}_k, \boldsymbol{\alpha}_m) = F[U_{\mathcal{X}_k}(\boldsymbol{\alpha}_k), U_{\mathcal{X}_m|\mathcal{X}_k}(\boldsymbol{\alpha}_m|\boldsymbol{\alpha}_k)]$$

and

$$U_{\mathcal{X}_m \mathcal{X}_k}(\boldsymbol{\alpha}_m, \boldsymbol{\alpha}_k) = F[U_{\mathcal{X}_m}(\boldsymbol{\alpha}_m), U_{\mathcal{X}_k|\mathcal{X}_m}(\boldsymbol{\alpha}_k|\boldsymbol{\alpha}_m)].$$

The aggregation is *exchangeable* if

$$F[U_{\mathcal{X}_k}(\boldsymbol{\alpha}_k), U_{\mathcal{X}_m|\mathcal{X}_k}(\boldsymbol{\alpha}_m|\boldsymbol{\alpha}_k)] = F[U_{\mathcal{X}_m}(\boldsymbol{\alpha}_m), U_{\mathcal{X}_k|\mathcal{X}_m}(\boldsymbol{\alpha}_k|\boldsymbol{\alpha}_m)].$$

□

Essentially, exchangeability requires that a notion of reciprocity exists. If not, then either (a) some information has been lost or ignored in the reframing, (b) the information has not been applied consistently, or (c) a condition of intransigence exists on the part of one or more agents, such that no reciprocity is possible. Obviously, the utilities of classical game theory are exchangeable, since then the conditional utilities coincide with the categorical utilities. Thus, exchangeability is a weaker condition than the assumption that all utilities are categorical.

The notion of exchangeability was first introduced in the probabilistic context by de Finetti (1937). Let $\{Y_1,\ldots,Y_n\}$ be a set of random variables and let $\{\pi_1,\ldots,\pi_n\}$ be an arbitrary permutation of $\{1,\ldots,n\}$. This set is said to be *exchangeable* if the joint distribution satisfies the condition $F_{Y_1\ldots Y_n}(y_1,\ldots,y_n) = F_{Y_{\pi_1}\ldots Y_{\pi_n}}(y_{\pi_1},\ldots,y_{\pi_n})$. Exchangeability means that the joint distribution of a set of random variables is independent of the order in which they are considered. Although often not mentioned in probabilistic discussions, exchangeability is central to the development of Bayesian statistical theory, sampling theory, and many other statistical inference operations. Indeed, we immediately see that exchangeability is satisfied with probability theory since, by Bayes rule,

$$p_{\Omega_1\Omega_2}(\omega_1,\omega_2) = p_{\Omega_1}(\omega_1)p_{\Omega_2|\Omega_1}(\omega_2|\omega_1) = p_{\Omega_2}(\omega_2)p_{\Omega_1|\Omega_2}(\omega_1|\omega_2)$$
$$= p_{\Omega_2\Omega_1}(\omega_2,\omega_1).$$

Thus, the two epistemological framings $\{p_{\Omega_1}, p_{\Omega_2|\Omega_1}\}$ and $\{p_{\Omega_2}, p_{\Omega_1|\Omega_2}\}$ are exchangeable.

To illustrate probabilistic exchangeability, let us return to the weather prediction problem discussed in Example 2.4. The way a statistical inference problem is framed certainly affects the way the information is presented for analysis, but it ought not affect the relevance of the information. If, instead of the original framing of the weather scenario but with the same set of information, we were to determine an epistemic ordering of storminess today and conditional epistemic orderings of cloudiness yesterday given storminess today and the conditional degree of belief regarding cloudiness yesterday given storminess today, the aggregated ordering of the joint weather state for the two days should yield the same result as obtained with the original framing. Exchangeability amounts to nothing more than invariance with respect to the way information is presented. If exchangeability were to fail, then some information is either not used or is used in an inconsistent way. Exchangeability is implicitly assumed whenever probability is used to model random phenomena. If it does not hold, then the use of probability to model random behavior is inappropriate.

In the praxeological context, if exchangeability does not hold, then at least one stakeholder is immune to social forces; that is, reciprocal relationships between that stakeholder and others simply do not exist. However, even under the doctrine of individual rationality, it is usually not assumed that the individuals form their preferences in a complete social vacuum, such that the preferences of others do not influence their own. Indeed, the field of behavioral economics, which retains the fundamental model of individual rationality, employs concepts such as inequity aversion, fairness, and reciprocity in the design of individual utility functions (Bolton and Ockenfels, 2005; Fehr and Schmidt, 1999; Sen, 1990; Camerer, 2003; Camerer et al., 2004b, 2004a; Henrich et al., 2005).

The notion of exchangeability can be somewhat controversial, since it requires that the concordant utility be independent of the way the problem is framed, so long as the same information is used. Let us examine this assumption from two points of view. First, consider the problem of using this approach to analyze human behavior. Fundamental psychological and sociological issues bring such a richness and variability to human decision making that it is difficult to define a model that accounts for more than a few somewhat idealized situations. We do not presume that exchangeability will apply to all social situations. Indeed, if one stakeholder were willing to compromise but the other were intransigent, then there would be no way to ensure that the problem can be reframed in a way that results in the same concordant utility. There is no universal model of social behavior. The endogenous aggregation hypothesis, however, is a legitimate generalization of the individual rationality hypothesis in the sense that a classical social welfare function is a function of, and only of, the individual self-interested utilities, and therefore trivially conforms to the exchangeability hypothesis.

Although the exchangeability principle may be controversial as a model of human behavior, its application to the design and synthesis of artificially intelligent systems is much less so. Artificially intelligent agents must operate according to the model that is used to design them. If they are designed to use all of the available information in a consistent way, then it is not unreasonable to assume that their preference orderings can be reframed and maintain exchangeability. Certainly, the designers of an artificially intelligent group, such as a group of robots who are designed to function harmoniously, could not object to such a principle. In fact, it would be highly desirable as a fundamental regularity property that would reduce or eliminate inconsistent or contradictory behavior.

The final principle necessary to define a synthesis procedure deals with the common sense notion that in the absence of opposition, the group must not arbitrarily override the wishes of the individual. Thus, if X_1 prefers a red convertible to a green sedan and X_2 is indifferent between the two alternatives, the group $\{X_1, X_2\}$ cannot prefer a green sedan to a red convertible. Accordingly, we offer the following principle:

Corollary 2.5 Principle 5 (Monotonicity) *If a subgroup prefers one alternative to another and the complementary subgroup is indifferent with respect* *to the two alternatives, then the group as a whole must not prefer the latter alternative to the former one.*

In terms of utilities, this principle requires that if one of the arguments of F is held constant and the other is increased, then the aggregated utility should not decrease. Thus, if

$$U_{x_m | x_k}(\alpha_m | \alpha_k) \geq U_{x_m | x_k}(\alpha'_m | \alpha'_k) \text{ and } U_{x_k}(\alpha_k) = U_{x_k}(\alpha'_k),$$

then

$$U_{x_m x_k}(\alpha_m, \alpha_k) \geq U_{x_m x_k}(\alpha'_m, \alpha'_k),$$

and if

$$U_{x_k}(\alpha_k) \geq U_{x_k}(\alpha'_k) \text{ and } U_{x_m x_k}(\alpha_m | \alpha_k) = U_{x_m x_k}(\alpha'_m | \alpha'_k),$$

then

$$U_{x_m x_k}(\alpha_m, \alpha_k) \geq U_{x_m x_k}(\alpha'_m, \alpha'_k).$$

These relationships require F to be nondecreasing in both arguments.

2.2.4 The aggregation theorem

The five aforementioned principles (conditioning, endogeny, acyclicity, exchangeability, and monotonicity) provide the framework within which to define an aggregation function that addresses the issue of how to combine preferences of individuals and small groups in a way that takes into account the social relationships that exist among them. The first two principles, conditioning and endogeny, are fundamental concepts that distinguish the present approach from classical decision theory and classical game theory. Within the framework provided by these principles, however, there are many possible aggregation structures. The last three principles (acyclicity, exchangeability, and monotonicity) represent operational and common sense concerns that are used to delimit the possibilities and permit us to focus on a specific aggregation procedure. The five principles, taken together, lead to the following notion of aggregation:

Definition 2.14 Given a multistakeholder decision problem in a social context, a family of concordant utilities is *endogenously aggregated* if there exists an aggregation function F that complies with the acyclicity, exchangeability, and monotonicity principles.   □

Theorem 2.1 (Aggregation Theorem) *Let* $\boldsymbol{\mathcal{X}}_n = \{X_1, \ldots, X_n\}$ *be an n-member multistakeholder group and let* $\boldsymbol{\mathcal{Q}}_m$ *denote the set of all m-element subgroups of* $\boldsymbol{\mathcal{X}}_n$. *That is,* $\boldsymbol{\mathcal{X}}_m \in \boldsymbol{\mathcal{Q}}_m$ *if* $\boldsymbol{\mathcal{X}}_m = \{X_{i_1}, \ldots, X_{i_m}\}$ *with* $1 \leq i_1 < \cdots < i_m \leq n$. *Let* $\{U_{\boldsymbol{\mathcal{X}}_m} : \boldsymbol{\mathcal{X}}_m \in \boldsymbol{\mathcal{Q}}_m, m = 1, \ldots, n\}$ *be a family of concordant utilities and let*

$$\{U_{\boldsymbol{\mathcal{X}}_m|\boldsymbol{\mathcal{X}}_k} : \boldsymbol{\mathcal{X}}_m \cap \boldsymbol{\mathcal{X}}_k = \varnothing, \boldsymbol{\mathcal{X}}_m \in \boldsymbol{\mathcal{Q}}_m, \boldsymbol{\mathcal{X}}_k \in \boldsymbol{\mathcal{Q}}_k, \ m + k \leq n\}$$

be a family of conditional concordant utilities associated with all pairs of disjoint subgroups of $\boldsymbol{\mathcal{X}}_n$. *These utilities are endogenously aggregated if and only if, for every pair of disjoint subgroups* $\boldsymbol{\mathcal{X}}_m$ *and* $\boldsymbol{\mathcal{X}}_k$,

$$U_{\boldsymbol{\mathcal{X}}_m \boldsymbol{\mathcal{X}}_k}(\boldsymbol{\alpha}_m, \boldsymbol{\alpha}_k) = F[U_{\boldsymbol{\mathcal{X}}_k}(\boldsymbol{\alpha}_k), U_{\boldsymbol{\mathcal{X}}_m \boldsymbol{\mathcal{X}}_k}(\boldsymbol{\alpha}_m | \boldsymbol{\alpha}_k)] = U_{\boldsymbol{\mathcal{X}}_m \boldsymbol{\mathcal{X}}_k}(\boldsymbol{\alpha}_m | \boldsymbol{\alpha}_k) U_{\boldsymbol{\mathcal{X}}_k}(\boldsymbol{\alpha}_k). \tag{2.5}$$

This theorem was originally introduced in the probabilistic context by Cox (1946) as an alternative development of probability theory, and is also discussed by Jaynes (2003) and Tribus (1969). Abbas (2003) has applied this result to utility theory. The proof below follows Jaynes (2003).

Proof of the aggregation theorem We first prove the general case $n \geq 3$, then specialize to the case $n = 2$. Let $\boldsymbol{\mathcal{X}}_i$, $\boldsymbol{\mathcal{X}}_j$, and $\boldsymbol{\mathcal{X}}_k$ be arbitrary pairwise disjoint subgroups of $\boldsymbol{\mathcal{X}}_n$, and let $U_{\boldsymbol{\mathcal{X}}_i \boldsymbol{\mathcal{X}}_j \boldsymbol{\mathcal{X}}_k}$, $U_{\boldsymbol{\mathcal{X}}_i | \boldsymbol{\mathcal{X}}_j \boldsymbol{\mathcal{X}}_k}$, $U_{\boldsymbol{\mathcal{X}}_i \boldsymbol{\mathcal{X}}_j | \boldsymbol{\mathcal{X}}_k}$, $U_{\boldsymbol{\mathcal{X}}_i \boldsymbol{\mathcal{X}}_j}$, $U_{\boldsymbol{\mathcal{X}}_i | \boldsymbol{\mathcal{X}}_j}$, and $U_{\boldsymbol{\mathcal{X}}_i}$ be endogenously aggregated concordant utilities. That is,

$$U_{\boldsymbol{\mathcal{X}}_i \boldsymbol{\mathcal{X}}_j \boldsymbol{\mathcal{X}}_k}(\boldsymbol{\alpha}_i, \boldsymbol{\alpha}_j, \boldsymbol{\alpha}_k) = F[U_{\boldsymbol{\mathcal{X}}_j \boldsymbol{\mathcal{X}}_k}(\boldsymbol{\alpha}_j, \boldsymbol{\alpha}_k), U_{\boldsymbol{\mathcal{X}}_i | \boldsymbol{\mathcal{X}}_j \boldsymbol{\mathcal{X}}_k}(\boldsymbol{\alpha}_i | \boldsymbol{\alpha}_j, \boldsymbol{\alpha}_k)] \tag{2.6}$$

$$= F[U_{\boldsymbol{\mathcal{X}}_k}(\boldsymbol{\alpha}_k), U_{\boldsymbol{\mathcal{X}}_i \boldsymbol{\mathcal{X}}_j | \boldsymbol{\mathcal{X}}_k}(\boldsymbol{\alpha}_i, \boldsymbol{\alpha}_j | \boldsymbol{\alpha}_k)]. \tag{2.7}$$

But

$$U_{\boldsymbol{\mathcal{X}}_j \boldsymbol{\mathcal{X}}_k}(\boldsymbol{\alpha}_j, \boldsymbol{\alpha}_k) = F[U_{\boldsymbol{\mathcal{X}}_k}(\boldsymbol{\alpha}_k), U_{\boldsymbol{\mathcal{X}}_j | \boldsymbol{\mathcal{X}}_k}(\boldsymbol{\alpha}_j | \boldsymbol{\alpha}_k)] \tag{2.8}$$

and

$$U_{\boldsymbol{\mathcal{X}}_i \boldsymbol{\mathcal{X}}_j | \boldsymbol{\mathcal{X}}_k}(\boldsymbol{\alpha}_i, \boldsymbol{\alpha}_j | \boldsymbol{\alpha}_k) = F[U_{\boldsymbol{\mathcal{X}}_j | \boldsymbol{\mathcal{X}}_k}(\boldsymbol{\alpha}_j | \boldsymbol{\alpha}_k), U_{\boldsymbol{\mathcal{X}}_i | \boldsymbol{\mathcal{X}}_j \boldsymbol{\mathcal{X}}_k}(\boldsymbol{\alpha}_i | \boldsymbol{\alpha}_j, \boldsymbol{\alpha}_k)]. \tag{2.9}$$

Substituting (2.8) into (2.6) and (2.9) into (2.7) yields

$$F\left[F[U_{\boldsymbol{\mathcal{X}}_k}(\boldsymbol{\alpha}_k), U_{\boldsymbol{\mathcal{X}}_j | \boldsymbol{\mathcal{X}}_k}(\boldsymbol{\alpha}_j | \boldsymbol{\alpha}_k)], U_{\boldsymbol{\mathcal{X}}_i | \boldsymbol{\mathcal{X}}_j \boldsymbol{\mathcal{X}}_k}(\boldsymbol{\alpha}_i | \boldsymbol{\alpha}_j, \boldsymbol{\alpha}_k) \right]$$

$$= F\left[U_{\boldsymbol{\mathcal{X}}_k}(\boldsymbol{\alpha}_k), F[U_{\boldsymbol{\mathcal{X}}_j | \boldsymbol{\mathcal{X}}_k}(\boldsymbol{\alpha}_j | \boldsymbol{\alpha}_k), U_{\boldsymbol{\mathcal{X}}_i | \boldsymbol{\mathcal{X}}_j \boldsymbol{\mathcal{X}}_k}(\boldsymbol{\alpha}_i | \boldsymbol{\alpha}_j, \boldsymbol{\alpha}_k)] \right]. \tag{2.10}$$

In terms of general arguments, this equation becomes

$$F[F(x,y),z] = F[x,F(y,z)],\qquad(2.11)$$

called the *associativity equation*, which has been studied extensively by Abel (1826) and Aczél (1966). By direct substitution, it is easy to see that (2.11) is satisfied if

$$f[F(x,y)] = f(x)f(y)\qquad(2.12)$$

for any function f. It has been shown by Cox (1946) that if F is differentiable in both arguments, then (2.12) is the general solution to (2.11). Taking f as the identity function, $F(x,y) = xy$, and

$$U_{\mathcal{X}_i\mathcal{X}_j}(\boldsymbol{\alpha}_i,\boldsymbol{\alpha}_j) = F\big[U_{\mathcal{X}_i}(\boldsymbol{\alpha}_i),U_{\mathcal{X}_j|\mathcal{X}_i}(\boldsymbol{\alpha}_j|\boldsymbol{\alpha}_i)\big]$$

$$= U_{\mathcal{X}_i}(\boldsymbol{\alpha}_i)U_{\mathcal{X}_j|\mathcal{X}_i}(\boldsymbol{\alpha}_j|\boldsymbol{\alpha}_i).$$

When $n = 2$, we may proceed by defining a dummy stakeholder X_3 with corresponding singleton action space $\mathcal{A}_3 = \{a_3\}$. We then define

$$U_{X_3}(\boldsymbol{\alpha}_3) = c\ \forall\,\boldsymbol{\alpha}_3$$

where c is chosen such that

$$\sum_{a_1,a_2} U_{X_3}(a_1,a_2,a_3) = 1,$$

and

$$U_{X_2|X_3}(\boldsymbol{\alpha}_2|\boldsymbol{\alpha}_3) = U_{X_2}(\boldsymbol{\alpha}_2)\ \forall\,\boldsymbol{\alpha}_3$$

$$U_{X_1|X_2X_3}(\boldsymbol{\alpha}_1|\boldsymbol{\alpha}_2\boldsymbol{\alpha}_3) = U_{X_1|X_2}(\boldsymbol{\alpha}_1|\boldsymbol{\alpha}_2)\ \forall\,\boldsymbol{\alpha}_3.$$

Inserting these expressions into (2.10) yields

$$F\Big[F\big[U_{X_3}(\boldsymbol{\alpha}_3),U_{X_2}(\boldsymbol{\alpha}_2)\big],U_{X_1|X_2}(\boldsymbol{\alpha}_1|\boldsymbol{\alpha}_2)\Big]$$

$$= F\Big[U_{X_3}(\boldsymbol{\alpha}_3),F\big[U_{X_2}(\boldsymbol{\alpha}_2),U_{X_1|X_2}(\boldsymbol{\alpha}_1|\boldsymbol{\alpha}_2)\big]\Big]$$

and the result follows.

To prove the converse, we note that F given by (2.5) is nondecreasing in both arguments since $U_{\mathcal{X}_k}$ and $U_{\mathcal{X}_m|\mathcal{X}_k}$ are nonnegative. Also, since the subgroups $\mathcal{X}_m$ and $\mathcal{X}_k$ are arbitrary, (2.5) holds if we reverse the roles of m and k. Thus, exchangeability is satisfied and the aggregation is endogenous. $\square$

The aggregation theorem establishes that, upon compliance with the afore-mentioned principles, endogenous utility aggregation in the praxeological domain conforms to the same mathematical syntax as does the aggregation of evidence in the epistemological domain, namely, the law of total probability. There is an apparent difference, however, in the structure of a concordant utility and a conventional multivariate probability mass function. A probability mass function for a family of k random variables is a function of k independent variables, but a concordant utility for k stakeholders, although it is a mass function (i.e., it is nonnegative and sums to unity), it is a function of nk independent variables. To appreciate this difference, we observe that, whereas a conventional multivariate mass function associates a scalar value with each random variable, a concordant utility associates an n-element action profile with each stakeholder.

To make this difference explicit, consider a collection of three random variables $\{Z_1, Z_2, Z_3\}$. The multivariate probability mass function $p_{Z_1 Z_2 Z_3}(z_1, z_2, z_3)$ is the probability that $Z_i = z_i$, $i = 1, 2, 3$. By contrast, the three-stakeholder concordant utility $U_{X_1 X_2 X_3}(\mathbf{a}_1, \mathbf{a}_2, \mathbf{a}_3)$ quantifies the degree of concordance that obtains if X_i conjectures $\mathbf{a}_i$, $i = 1, 2, 3$.

In Appendix A we show that the appropriate probabilistic analog for a concordant utility is the joint distribution of a k-member family of n-dimensional random vectors. Consequently, a concordant utility possesses all of the syntactical properties commonly associated with probability mass functions. including independence, the chain rule, marginalization, and Bayes rule, albeit with praxeological semantics.

Independence Let $\mathcal{X}_m$ and $\mathcal{X}_k$ be disjoint subgroups of $\mathcal{X}_n$. These subgroups are *praxeologically independent* if neither subgroup influences the other; that is, if

$$U_{\mathcal{X}_m \mathcal{X}_k}(\boldsymbol{\alpha}_m, \boldsymbol{\alpha}_k) = U_{\mathcal{X}_m}(\boldsymbol{\alpha}_m) U_{\mathcal{X}_k}(\boldsymbol{\alpha}_k)$$

or, equivalently,

$$U_{\mathcal{X}_m}(\boldsymbol{\alpha}_m) = U_{\mathcal{X}_m | \mathcal{X}_k}(\boldsymbol{\alpha}_m | \boldsymbol{\alpha}_k)$$

for all $\boldsymbol{\alpha}_k$.

The Chain Rule Let $\mathcal{X}_m$, $\mathcal{X}_k$, and $\mathcal{X}_\ell$ be pairwise disjoint subgroups of $\mathcal{X}_n$. Then

$$U_{\mathcal{X}_m \mathcal{X}_k \mathcal{X}_\ell}(\boldsymbol{\alpha}_m, \boldsymbol{\alpha}_k, \boldsymbol{\alpha}_\ell) = U_{\mathcal{X}_m | \mathcal{X}_k \mathcal{X}_\ell}(\boldsymbol{\alpha}_m | \boldsymbol{\alpha}_k, \boldsymbol{\alpha}_\ell) U_{\mathcal{X}_k | \mathcal{X}_\ell}(\boldsymbol{\alpha}_k | \boldsymbol{\alpha}_l) U_{\mathcal{X}_\ell}(\boldsymbol{\alpha}_\ell).$$

The chain rule is the mechanism by which individual utilities can be aggregated to synthesize the concordant utility. Let us first recall that the acyclicity principle ensures that at least one stakeholder possesses a categorical utility. Without loss of generality, let us assume this condition holds for X_1. Applying the chain rule,

$$U_{X_1\cdots X_n}(\mathbf{a}_1,\ldots,\mathbf{a}_n) = u_{X_n|X_1\cdots X_{n-1}}(\mathbf{a}_n|\mathbf{a}_1,\ldots,\mathbf{a}_{n-1})U_{X_1\cdots X_{n-1}}(\mathbf{a}_1,\ldots,\mathbf{a}_{n-1})$$

and

$$U_{X_1\cdots X_{n-1}}(\mathbf{a}_1,\ldots,\mathbf{a}_{n-1})$$
$$= u_{X_{n-1}|X_1\cdots X_{n-2}}(\mathbf{a}_{n-1}|\mathbf{a}_1\ldots,\mathbf{a}_{n-2})U_{X_1\cdots X_{n-2}}(\mathbf{a}_1,\ldots,\mathbf{a}_{n-2}),$$

continuing this process until exhaustion and substituting, we obtain

$$U_{X_1\cdots X_n}(\mathbf{a}_1,\ldots,\mathbf{a}_n) = u_{X_n|X_1\cdots X_{n-1}}(\mathbf{a}_n|\mathbf{a}_1,\ldots,\mathbf{a}_{n-1})$$
$$u_{X_{n-1}|X_1\cdots X_{n-2}}(\mathbf{a}_{n-1}|\mathbf{a}_1\ldots,\mathbf{a}_{n-2})\ldots u_{X_1}(\mathbf{a}_1). \quad (2.13)$$

The utilities on the right-hand side of (2.13) are termed *ex ante utilities*, meaning that they are defined before the game is played.

Marginalization Let $\mathcal{X}_m = \{X_{j_1},\ldots,X_{j_m}\}$ and $\mathcal{X}_k = \{X_{i_1},\ldots,X_{i_k}\}$ be disjoint subgroups of $\mathcal{X}_n$. Then the marginal concordant utility of $\mathcal{X}_m$ with respect to the subgroup $\{\mathcal{X}_m,\mathcal{X}_k\}$ is obtained by summing over $\mathcal{A}^k$, yielding

$$U_{\mathcal{X}_m}(\boldsymbol{\alpha}_m) = \sum_{\boldsymbol{\alpha}_k} U_{\mathcal{X}_m \mathcal{X}_k}(\boldsymbol{\alpha}_m,\boldsymbol{\alpha}_k),$$

and the marginal utility of the individual stakeholder X_i is given by

$$u_{X_i}(\mathbf{a}_i) = \sum_{\sim\mathbf{a}_i} U_{\mathcal{X}_n}(\mathbf{a}_1,\ldots,\mathbf{a}_n), \quad (2.14)$$

where the notation $\sum_{\sim\mathbf{a}_i}$ means that the sum is taken over all arguments except $\mathbf{a}_i$.

Marginalization provides the mechanism for individual preferences to emerge as a result of the social relationships that can exist between individuals. Thus, even though an ex ante categorical ordering may not be given, marginalization provides an ex post unconditional ordering; that is, after consideration of the social relationships among the stakeholders have been taken into account.

Since marginal utilities are not conditional, they may be used in accordance with classical solution concepts, such as equilibria, to solve the decision problem. Rather than proceed in this fashion, however, we defer further discussion of solution concepts to Chapter 3, where we shall see that, in addition to classical solution concepts that do not explicitly account for group-level interests, we can identify solution concepts that simultaneously accommodate both group and individual interests.

Example 2.5 Let $\{X_1, X_2, X_3\}$ comprise a three-agent group such that X_1 possesses a categorical utility u_{X_1}, X_2 possesses a conditional utility $u_{X_2|X_1}$, and X_3 possesses a conditional utility $u_{X_3|X_1X_2}$. Applying the chain rule yields the concordant utility

$$U_{X_1X_2X_3}(\mathbf{a}_1, \mathbf{a}_2, \mathbf{a}_3) = u_{X_1}(\mathbf{a}_1)u_{X_2|X_1}(\mathbf{a}_2|\mathbf{a}_1)u_{X_3|X_1X_2}(\mathbf{a}_3|\mathbf{a}_1, \mathbf{a}_2).$$

Since X_1 possesses an ex ante categorical utility, the ex post utility for X_1 will coincide with the categorical utility. The ex post utility of X_2 is given by

$$u_{X_2}(\mathbf{a}_2) = \sum_{\mathbf{a}_1}\sum_{\mathbf{a}_3}U_{X_1X_2X_3}(\mathbf{a}_1, \mathbf{a}_2, \mathbf{a}_3),$$

with a similar expression for the ex post utility for X_3. $\qquad\square$

Bayes Rule Since all of the mathematical properties of probability theory apply to utilities that comply with the aggregation theorem, we may apply Bayes rule. Let $\mathcal{X}_m$ and $\mathcal{X}_k$ be disjoint subgroups of $\mathcal{X}_n$, and suppose $U_{\mathcal{X}_k}$ and $U_{\mathcal{X}_m|\mathcal{X}_k}$ are given. Then

$$U_{\mathcal{X}_k|\mathcal{X}_m}(\boldsymbol{\alpha}_k|\boldsymbol{\alpha}_m) = \frac{U_{\mathcal{X}_m|\mathcal{X}_k}(\boldsymbol{\alpha}_m|\boldsymbol{\alpha}_k)U_{\mathcal{X}_k}(\boldsymbol{\alpha}_k)}{\sum_\eta U_{\mathcal{X}_m|\mathcal{X}_k}(\boldsymbol{\alpha}_m|\eta)U_{\mathcal{X}_k}(\eta)}.$$

Clearly, Bayes rule is a manifestation of exchangeability and provides a mechanism for reframing a multistakeholder decision problem.

The application of the probabilistic syntax to the structure and aggregation of utilities is not without controversy. One potentially controversial issue is normalization. This constraint imposes an important requirement on the utility, one that is already familiar to probabilists; namely, probability theory does not permit agnosticism – one cannot withhold belief from a proposition without ascribing the belief to its complement. That is, if $Prob(A) = \epsilon$, then $Prob(A^c) = 1 - \epsilon$. Similarly, utilities do not permit abstention – one is obligated to apportion one's entire unit of utility among the alternatives. This is exactly the issue raised by Shafer (1976) as a fundamental problem with the law of total probability.[6] However, there is an important difference between the epistemic and the praxeic contexts. In the epistemic context one can withhold committing to a belief, but in the praxeic context, one cannot withhold

---

[6] The so-called Dempster-Shafer theory is an alternative to classical probability theory. It holds that probability theory cannot distinguish between uncertainty and ignorance, since it does not permit one to withhold belief from a proposition without ascribing the belief to its complement. Dempster-Shafer theory is an interesting alternative to probability theory, especially when subjective considerations dominate to the extent that the agent is simply unable to assign degrees of belief to all possibilities.

taking action. Even if the problem context permits choosing the null action (do nothing), consequences will result.

2.3 Coherence

One of the important reasons why probability theory is a widely used model is because it is the only model of belief formulation by which, if an agent acts according to its beliefs, it will behave coherently; that is, it will never take an action that puts it at a categorical disadvantage such that no matter what action it would take, it would be worse off for having taken it. This property is the content of the Dutch Book theorem. A Dutch book is a gambling situation where the gambler cannot avoid sure loss – no matter what the outcome, the gambler suffers a net loss (its reward will always be less than its stake).

To illustrate the concept of sure loss, let r and d denote weather conditions of rain and dry, respectively, and let h and l denote sky conditions of heavy clouds and light clouds, respectively. Let $\succsim_w$ and $\succsim_s$ denote plausibility preference orderings for the weather and sky conditions, respectively, and let $\succsim_{ws}$ denote the plausibility preference ordering for the joint weather–sky conditions. Suppose $(r,h) \succsim_{ws} (d,h)$ and $(r,l) \succsim_{ws} (d,l)$. To avoid a Dutch book, $r \succsim_w d$ must hold. To see why this must be so, suppose you were to enter a lottery to guess whether or not it will rain, with the possibility to win a prize of \$1 for a correct guess. One concept of a fair entry fee would be \$$p$ (where $0 \le p \le 1$), the plausibility of $(r,h) \vee (r,l)$. Another concept of a fair entry fee would be \$$q$ (where $0 \le q \le 1$), the plausibility of d. By paying an entry fee of $p + q$, you would be assured of winning \$1, regardless of the outcome. Now suppose you consider the disjunction $(r,h) \vee (r,l)$ to be at least as plausible as the disjunction $(d,h) \vee (d,l)$ but you also consider d to be at least as plausible as r. According to these two plausibility orderings, a fair entry fee would require $p \ge \frac{1}{2}$ and $q \ge \frac{1}{2}$. Thus, it would be fair, in your estimation, to pay $p + q \ge 1$ to return \$1 for certain, resulting in a sure loss of $p + q - 1$. If you were to reason this way, you would be the victim of a Dutch book: Your behavior would not be coherent.

To be coherent, if you were to believe $(r,h) \vee (r,l)$ to be at least as plausible as $(d,h) \vee (d,l)$ (i.e., weather conditions involving rain, regardless of the cloud conditions, are always at least as plausible as weather conditions involving no rain), then you must also believe that rain is more plausible than no rain.

More generally, let $\Omega = \Omega_1 \times \cdots \times \Omega_n$ denote a product sample space with elements $\omega = (\omega_1, \ldots, \omega_n)$. Let $\succsim_\Omega$ be a plausibility ordering over Ω, and let $\succsim_{\Omega_i}$ be a plausibility ordering over Ω_i for $i = 1, \ldots, n$. To avoid sure loss, it must hold that if

$$(\omega_1,\ldots,\omega_i,\ldots,\omega_n) \succsim_{\Omega} (\omega_1,\ldots,\omega_i',\ldots,\omega_n)$$

for all subvectors $(\omega_1,\ldots,\omega_{i-1},\omega_{i+1},\ldots,\omega_n)$, then it must hold that

$$\omega_i \succsim_{\Omega_i} \omega_i'.$$

To establish this claim, let us consider the contrapositive. Suppose

$$\omega_i' \succ_{\Omega_i} \omega_i \qquad (2.15)$$

but there exists no subvector $(\omega_1,\ldots,\omega_{i-1},\omega_{i+1},\ldots,\omega_n)$ such that

$$(\omega_1,\ldots,\omega_{i-1},\omega_i',\omega_{i+1},\ldots,\omega_n) \succ_{\Omega} (\omega_1,\ldots,\omega_{i-1},\omega_i,\omega_{i+1},\ldots,\omega_n). \qquad (2.16)$$

This situation would invite a Dutch book, since if one were to enter a lottery to earn \$1 if ω' is true, a fair entry fee based on (2.15) would be $q > \frac{1}{2}$, since, viewed in isolation, ω_i' is more plausible ω_i. However, a fair entry fee based on the negation of (2.16) would be $p > \frac{1}{2}$ since there exists no subvector $(\omega_1,\ldots,\omega_{i-1},\omega_{i+1},\ldots,\omega_n)$ such that, when considered jointly with ω_i and ω_i', the vector that incorporates ω_i is at least as plausible as the vector that incorporates ω_i'. Thus, if a Dutch book is not possible, then there must exist at least one subvector $(\omega_1^*,\ldots,\omega_{i-1}^*,\omega_{i+1}^*,\ldots,\omega_n^*)$ such that

$$(\omega_1^*,\ldots,\omega_{i-1}^*,\omega_i,\omega_{i+1}^*,\ldots,\omega_n^*) \succsim_{\Omega} (\omega_1^*,\ldots,\omega_{i-1}^*,\omega_i',\omega_{i+1}^*,\ldots,\omega_n^*).$$

The Dutch book theorem (Ramsey, 1950; de Finetti, 1937) and its converse (Kemeny, 1955) state that it is not possible to construct a Dutch book if and only if the gambler acts in accordance with a probability measure that describes the individual's degrees of plausibility regarding the propositions under consideration.

The fact that endogenous aggregation requires utilities to possess the mathematical structure of probability mass functions means that the notion of coherence also applies in the praxeological case. In particular, coherence means that no stakeholder will be categorically disenfranchised; that is, no stakeholder will be placed in a situation such that no choice that is good for the group will be good for it. Accordingly, if $\mathbf{a}$ is myopically preferred by X_i to $\mathbf{a}'$, that is, $\mathbf{a}_i|\varnothing \succsim_{X_i|\varnothing} \mathbf{a}_i'|\varnothing$, then there must be a joint conjecture for the subgroup $\sim X_i = \mathcal{X}_n \setminus \{X_i\}$ (the entire group with X_i removed) of the form $\{\mathbf{a}_1^*,\ldots,\mathbf{a}_{i-1}^*,\mathbf{a}_{i+1}^*,\ldots,\mathbf{a}_n^*\}$ such that

$$\{\mathbf{a}_1^*,\ldots,\mathbf{a}_{i-1}^*,\mathbf{a}_i,\mathbf{a}_{i+1}^*,\ldots,\mathbf{a}_n^*\} \unrhd_{\sim X_i} \{\mathbf{a}_1^*,\ldots,\mathbf{a}_{i-1}^*,\mathbf{a}_i',\mathbf{a}_{i+1}^*,\ldots,\mathbf{a}_n^*\}.$$

Coherence is a weak notion of equity that requires only that a group must allow for the possibility (but not the guarantee) that each individual can achieve its optimum, and is tantamount to avoiding *categorical subjugation*, whereby an individual stakeholder would be required to sacrifice its performance under all circumstances in order to benefit the larger group. This feature is perhaps a minimal condition for serious negotiations to exist. Essentially, every stakeholder has a seat at the table.

Categorical subjugation is similar to the notion of *suppression* as discussed by Hansson (1972) and Fishburn (1973). An individual is suppressed if, whenever it prefers alternative **a** to **a′**, then **a′** is preferred to **a** by the group.

Corollary 2.6 *Endogenous aggregation and coherence are equivalent concepts.*

Proof: The aggregation theorem establishes that endogenous aggregation obtains if and only if the utilities are mass functions with respect to a praxeological measure P_{χ_n}. Also, the Dutch book theorem and its converse establish that coherence obtains if and only if the utilities are mass functions corresponding to a praxeological measure.   □

This corollary establishes that an alternative principle to the endogeny principle is the following:

Corollary 2.7 Principle 2′ (Praxeological Coherence) *The interests of no individual should be categorically subjugated to the interests of the group in all situations.*

2.4 Utility networks

The analogy between the epistemological notion of probability and the praxeological notion of utility makes it possible to apply powerful graphical techniques to the study of utility theory. A graph is a mathematical structure that provides a pictorial representation of the pairwise social influence interrelationships between the elements of a group. When applied to multistakeholder decision making in the presence of dependencies, graph theory can enhance understanding and provide a visual summary of the overall connectivity of the stakeholders.

Graph theory is perhaps most useful, however, when the social influence connections are relatively sparse. Many configurations of stakeholders will primarily involve local relationships (spatially or functionally). The classical example of such influence relationships are joint probability mass functions that involve conditional independencies. When this occurs, it is convenient to synthesize the joint distribution from conditional and marginal distributions. As Pearl (1988) observed, "Probabilistic judgments on a small number of propositions are issued swiftly and reliably, while judging the likelihood of a conjunction of propositions entails much difficulty and hesitancy. This suggests that the elementary building blocks of human knowledge are not entries of a joint-distribution table. Rather, they are low-order marginal and conditional probabilities defined over small clusters of propositions" (p. 78).

Pearl's (1988) observation also applies to the praxeological context by replacing "probability" with "utility." Accordingly, we might paraphrase Pearl's remark as follows: The elementary building blocks of a multistakeholder group are low-order categorical and conditional utilities defined over small clusters of agents. Many organizational structures possess hierarchical components arranged into clusters, such as branches, departments, and teams, that constrain the flow of social influence (see Burns and Stalker, 1961).

When constructing a concordant utility, graph theory represents a powerful tool for understanding and manipulating the interrelationships between the individual stakeholders. Since the utilities possess the same mathematical structure as probability mass functions, we may apply graph theory to multistakeholder problems as well as to probabilistic (epistemic) problems. To develop this theory, however, we must introduce some basic graph-theoretic definitions and principles (for a more detailed account, see, for example, Cowell et al., 1999; Lauritzen, 1996; and Jensen, 2001).

A *graph* $\mathcal{G}$ is a pair $\mathcal{G} = (\boldsymbol{\mathcal{X}}_n, E)$, where $\boldsymbol{\mathcal{X}}_n$ is a finite set of *vertices*, of $\mathcal{G}$, also called *nodes*, and E is a subgroup of the product set $\boldsymbol{\mathcal{X}}_n \times \boldsymbol{\mathcal{X}}_n$ of ordered pairs of vertices, called the *edges* or *links* of $\mathcal{G}$. If X_i and X_j are vertices, that is, $X_i, X_j \in \boldsymbol{\mathcal{X}}_n$, then the ordered pair (X_i, X_j) is an edge if there is a connection from X_i to X_j, and we write $(X_i, X_j) \in E$. If the connection is two-way – that is, there is also connection from X_j to X_i – then the edge is said to be *undirected*, and we write $X_i \sim X_j$. But if the connection from X_i to X_j is one-way, the edge is said to be *directed*, and we write $X_i \to X_j$. Thus, if $(X_i, X_j) \in E$ is directed, then $(X_j, X_i) \notin E$. If all of the edges of a graph $\mathcal{G}$ are directed, then $\mathcal{G}$ is said to be a *directed graph*.

Let $\mathcal{G}$ be a graph with n vertices, and let $\{i_1, \ldots, i_k\}$ be a subset of $\{1, \ldots, n\}$. A *path* of length k from vertex X_{i_1} to vertex X_{i_k} is a sequence $\{X_{i_1}, \ldots, X_{i_k}\}$ of distinct vertices such that $(X_{i_j}, X_{i_{j+1}}) \in E$ for $j = 1, \ldots, k - 1$. A path never crosses itself, and if all of the edges along a path are directed, then it is a *directed path*. We write $X_i \mapsto X_j$ if there is a path from X_i to X_j.

The *ancestors* of X_i, denoted an (X_i), comprise all vertices X such that $X \mapsto X_i$, and the *descendants* of X_i, denoted de (X_i), comprise all vertices X such that $X_i \mapsto X$ but $X \not\mapsto X_i$. The set of nonancestors of X_i is defined as na $(X_i) = \boldsymbol{\mathcal{X}}_n \setminus (\text{an}(X_i) \cup \{X_i\})$, and the set of nondescendants of X_i is defined as nd $(X_i) = \boldsymbol{\mathcal{X}}_n \setminus (\text{de}(X_i) \cup \{X_i\})$.

If $X_i \to X_j$, then X_i is called a *parent* of X_j, and X_j is a *child* of X_i. The set of parents of X_i is denoted pa (X_i), and the set of *children* of X_i is

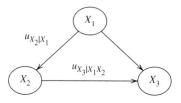

Figure 2.2. The DAG for a three-agent group.

denoted ch (X_i). (Note: These relationships pertain only to vertices connected by directed edges.) If a vertex has no parents, it is called a *root vertex*.

A *k-cycle* is a sequence $X_{i_1}, \ldots, X_{i_k}$ of vertices such that $(X_{i_j}, X_{i_{j+1}}) \in E$ for $j = 1, \ldots, k-1$, with the condition that all vertices are distinct with the exception that $X_{i_1} = X_{i_k}$. A directed graph with no cycles is said to be a *directed acyclic graph*, or DAG.

As an example, consider the DAG displayed in Figure 2.2, which corresponds to the three-agent group defined in Example 2.5. We see that pa$(X_1) = \varnothing$, pa$(X_2) = \{X_1\}$, and pa$(X_3) = \{X_1, X_2\}$.

Since, as Theorem 2.1 establishes, normalized utilities that comply with the principles of acyclicity, exchangeability, and monotonicity are mass functions, we may view the corresponding DAG as a Bayesian network, albeit with different semantics that usually accompany such graphs. Bayesian networks are well known in the literature (see, for example, Cowell et al., 1999; Lauritzen, 1996; and Jensen, 2001), with the vertices corresponding to random variables and the edges corresponding to conditional probability mass functions. With the praxeological context, however, since the nodes are stakeholders and the edges are conditional utilities that possess the mathematical syntax of probability mass functions, we shall term them *utility networks* (also, see Abbas, 2009). The key property of a DAG is the *Markov condition*: Nondescendent nonparents of a vertex have no influence on the vertex, given the state of its parent vertices (Cozman, 2000). Consequently, if a group can be represented as a DAG, the conditional utility of an agent is dependent only upon the conjectures of its parents, and the joint mass function of the vertices is the product of the conditional mass functions of all nonroot vertices and the marginal mass functions of all root vertices (Pearl, 1988; Cowell et al., 1999). Thus, using (2.13), the concordant utility of the joint conjecture $\{\mathbf{a}_1 \ldots, \mathbf{a}_n\}$ for the group $\mathcal{X}_n = \{X_1, \ldots, X_n\}$, becomes

$$U_{X_1 \cdots X_n}(\mathbf{a}_1, \ldots, \mathbf{a}_n) = \prod_{i=1}^{n} u_{X_i \mid \text{pa}(X_i)}(\mathbf{a}_i \mid \mathbf{a}_{i_1}, \ldots, \mathbf{a}_{i_{p_i}}), \qquad (2.17)$$

where $\{\mathbf{a}_{i_1}, \ldots, \mathbf{a}_{i_{p_i}}\}$ is the joint conjecture corresponding to $\mathrm{pa}(X_i) = \{X_{i_1}, \ldots, X_{i_{p_i}}\}$, the parents of X_i. If $p_i = 0$, then $u_{X_i|\mathrm{pa}(X_i)} = u_{X_i}$, a categorical utility. In particular, the concordant utility of the system defined is Figure 2.2 is

$$U_{X_1 X_2 X_3}(\mathbf{a}_1, \mathbf{a}_2, \mathbf{a}_3) = u_{X_1}(\mathbf{a}_1) u_{X_2|X_1}(\mathbf{a}_2|\mathbf{a}_1) u_{X_3|X_1 X_2}(\mathbf{a}_3|\mathbf{a}_1, \mathbf{a}_2).$$

When defining the relationships among the members of a group, a natural way to present the problem is to examine each stakeholder and determine the subgroup that influences it, thereby defining its parents. Once these individual conditional utilities are defined, aggregation to create the concordant utility can be achieved by applying (2.17). This equation is, therefore, the key result that establishes how to aggregate the ex ante utilities of a group.

2.5 Reframing

As mentioned in Section 2.2.4, the acyclicity constraint reduces the generality of the model, since it restricts social influence to a hierarchical flow. For the group modeled by the DAG illustrated in Figure 2.2, we see that X_1 directly influences both X_2 and X_3, but neither of those stakeholders directly influences X_1. Also, we see that X_2 directly influences X_3, but X_3 does not directly influence X_2. While it is true that the DAG structure does not permit cyclical direct influence, that does not mean that X_1 is not influenced by X_2 and X_3 in any way whatsoever. Let us pose the following question: Given that X_2 conjectures $\mathbf{a}_2$, what is the effect, if any, on X_1's preference for $\mathbf{a}_1$? In other words, can we define a conditional utility of the form $u_{X_1|X_2}(\mathbf{a}_1|\mathbf{a}_2)$? The answer is affirmative, and follows from Bayes rule:

$$u_{X_1|X_2}(\mathbf{a}_1|\mathbf{a}_2) = \frac{u_{X_2|X_1}(\mathbf{a}_2|\mathbf{a}_1) u_{X_1}(\mathbf{a}_1)}{u_{X_2}(\mathbf{a}_2)},$$

where $u_{X_2}(\mathbf{a}_2)$ is obtained via marginalization:

$$u_{X_2}(\mathbf{a}_2) = \sum_{\mathbf{a}_1, \mathbf{a}_3} U_{X_1 X_2 X_3}(\mathbf{a}_1, \mathbf{a}_2, \mathbf{a}_3).$$

We may also determine the social influence X_2 and X_3 jointly exert on X_1 by computing

$$u_{X_1|X_2 X_3}(\mathbf{a}_1|\mathbf{a}_2, \mathbf{a}_3) = \frac{U_{X_1 X_2 X_3}(\mathbf{a}_1, \mathbf{a}_2, \mathbf{a}_3)}{U_{X_2 X_3}(\mathbf{a}_2, \mathbf{a}_3)},$$

where

$$U_{X_2 X_3}(\mathbf{a}_2, \mathbf{a}_3) = \sum_{\mathbf{a}_1} U_{X_1 X_2 X_3}(\mathbf{a}_1, \mathbf{a}_2, \mathbf{a}_3).$$

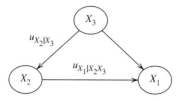

Figure 2.3. The DAG for the reframed three-agent group.

The social influence that X_3 exerts on X_2 is defined by

$$u_{X_2|X_3}(\mathbf{a}_2|\mathbf{a}_3) = \frac{U_{X_2X_3}(\mathbf{a}_2,\mathbf{a}_3)}{u_{X_3}(\mathbf{a}_3)},$$

where

$$u_{X_3}(\mathbf{a}_3) = \sum_{\mathbf{a}_1,\mathbf{a}_2} U_{X_1X_2X_3}(\mathbf{a}_1,\mathbf{a}_2,\mathbf{a}_3).$$

Using these calculations, we may reframe the group displayed in Figure 2.2 as

$$U_{X_1X_2X_3}(\mathbf{a}_1,\mathbf{a}_2,\mathbf{a}_3) = u_{X_1|X_2X_3}(\mathbf{a}_1|\mathbf{a}_2,\mathbf{a}_3)u_{X_2|X_3}(\mathbf{a}_2|\mathbf{a}_3)u_{X_3}(\mathbf{a}_3),$$

whose DAG is displayed in Figure 2.3, using the utilities as computed from the concordant utility as expressed with the preceding formulas. We assume, for all of these calculations, that all utilities are nonzero.

As originally framed, the group depicted in Figure 2.2 casts X_1 as the stakeholder at the top of the hierarchy, with X_2 subordinate to X_1 and X_3 subordinate to both X_1 and X_2. The reframed version of this problem reverses this hierarchy and puts X_3 as the top and X_1 at the bottom. It can readily be seen that other hierarchical structures are possible by computing the appropriate marginal and conditional utilities. The point of this exercise is that once the concordant utility is defined, many hierarchical structures can be defined that are compatible with it.

This example illustrates the consequences of the exchangeability principle. If a society complies with this principle, the social bonds that exist among its members enjoy an important notion of reciprocity such that if X_3 were at the top of the hierarchy and thus possessed a categorical utility, then the utilities conditioned on X_3's conjectures would reflect exactly the same social relationships, albeit expressed in a different way, as do the conditional utilities associated with the original framing. In other words, the social model is framing invariant.

Although the two framings represented by Figures 2.2 and 2.3 are both legitimate representations of the group, it is important to understand that the utilities used to characterize the second framing (Figure 2.3) are derived from the utilities used to characterize the original framing. The utilities u_{X_1}, $u_{X_2|X_1}$, and $u_{X_3|X_1X_2}$ are ex ante utilities – that is, utilities defined before social interaction takes place – but the utilities u_{X_3}, $u_{X_2|X_3}$, and $u_{X_1|X_2X_3}$ are ex post utilities that are defined as a result of the social interaction.

It must be pointed out, however, that under the exchangeability principle, if X_3 were originally at the top of the hierarchy, then the utilities associated with the framing illustrated in Figure 2.3 would be viewed as ex ante utilities, and the utilities associated with Figure 2.2 would be viewed as ex post utilities, being derived according to Bayes rule.

Reframing is an illustration of the power of mathematical models. Once the fundamental relationships are defined, the model can be manipulated in many ways to expose features that would otherwise be difficult to ascertain, and may facilitate the design and construction of artificially intelligent multiagent systems.[7] Reframing also permits a system designer to test the validity of the model by viewing it from different points of view.

2.6 Summary

We have presented a principle-based development of a new preference structure that is a true generalization of classical preference relations. Starting with an extension of ordinal relationships to include conditional preferences, we then introduce cardinal preference relations that permit stakeholders to modulate the strength of their preferences. This new preference structure is designed to accommodate the first two principles of our development, namely, conditioning and endogeny, which provide the framework to enable stakeholders to (a) expand beyond their own self-interest and (b) accommodate a notion of group preference as an emergent phenomenon that obtains as the conditional preferences propagate through the group.

The remaining three principles, namely, acyclicity, exchangeability, and monotonicity, are common sense and useful operational restrictions that enable the utilities to possess a convenient and powerful mathematical structure: the syntax of probability theory. Consequently, all of the familiar properties of probability theory, such as marginalization, independence, the chain rule, Bayes rule, and the Dutch book theorem, can be reinterpreted in a praxeological context.

---

[7] Such mathematical manipulations are common in engineering applications. For example, for a linear time-invariant system comprising the series connection of two simpler systems, the commutativity of convolution permits the order of the series connection to be reversed without changing the system transfer function. Such mathematical manipulations provide engineers the insight to design and build such systems with a minimal number of hardware components.

Two major benefits devolve from this structure. The first is that it provides stakeholders a natural mechanism to extend their spheres of interest beyond the self by adjusting their preferences as a function of the preferences of others. The second is that it permits a powerful graphical representation of a group such that the stakeholders are viewed as nodes and the conditional utilities are viewed as edges. This structure thus constitutes an exact praxeological analogue to the more familiar epistemological concept of a Bayesian network, complete with all of the intuition and functionality of such mathematical quantities.

3

Solutions

Any good mathematical commodity is worth generalizing.
— *Michael Spivak*
   Calculus on Manifolds (W. A. Benjamin, 1965)

The introduction of conditional utilities as a mechanism for modeling complex social behavior among multiple stakeholders raises the possibility of developing new notions of rational behavior that are consistent with this structure. Whereas individual rationality is the dominant behavioral concept with categorical utilities, the existence of conditional utilities makes it possible to extend beyond individual interests and to consider notions of rationality that account for both individual and group interests. We begin our study with a formal definition of a conditional game.

Definition 3.1 A *conditional game* is a triple $\{\mathcal{X}_n, \mathcal{A}, U_{\mathcal{X}_n}\}$, where $\mathcal{X}_n = \{X_1, \dots, X_n\}$ is a group of n stakeholders with product action space $\mathcal{A} = \mathcal{A}_1 \times \cdots \times \mathcal{A}_n$ and $U_{\mathcal{X}_n} = U_{X_1 \cdots X_n}$ is a concordant utility. Equivalently, by application of (2.17), a conditional game can be defined in terms of the conditional utilities $u_{X_i|\mathrm{pa}(X_i)}, i = 1, \dots, n$. If all utilities are categorical, a conditional game becomes a classical game. $\quad\square$

Once a game has been defined in terms of the players, actions, and payoffs, it remains to solve it. Each player must identify action profiles that are not only good for it but good for others and, therefore, likely to be instantiated. This is a nontrivial endeavor. Even if the player is motivated to do the best for itself, the fact that the actions of other players, which are not known in advance, will affect its payoff means that the player cannot simply maximize its payoff as if isolated from the effects of others. This realization is the basis for searching for an equilibrium. Of course, there will exist special games for which it is possible for a player not only to define what outcome is best for itself but also to ensure that that outcome actually obtains, but such situations are rare and usually not very theoretically interesting.

A *solution concept* is a logical rule that forms the basis for a player to predict how others might rationally behave and to act accordingly. As described by Shubik (1982), "Each solution probes some particular aspect of societal rationality, that is, the possible, proposed, or predicted behavior of rational individuals in mutual interaction" (p. 2).

As was discussed in Chapter 1, if all utilities are categorical, then individual rationality is the only operative philosophical basis for forming solutions, since defining a meaningful notion of group rationality is problematic. As developed in Chapter 2, however, the concordant utility contains a complete characterization of all social relationships that emerge as the conditional preferences of the individuals propagate through the group. Consequently, we investigate a number of solution concepts that extend interest beyond the self. We first define a solution concept that may be applied prior to aggregation, resulting in a conditioned Nash equilibrium. We next consider a solution concept defined at the group level; that is, a solution that accounts for the interests of the group as a whole, but which does not explicitly indicate the benefit to the individuals. We then consider how to extract individual ex post unconditional (categorical) utilities from the concordant utility. Once these are defined, we may then implement the standard solution concepts based on individual rationality, such as Nash equilibrium. Finally, we consider solution concepts that simultaneously take into consideration the interests of the group and the individuals, and develop group and individual welfare functions.

3.1 Conditioned Nash equilibria

Recall that a Nash equilibrium is an action profile such that, if any single agent were to change its action, it would reduce its payoff. We may easily extend this notion to the conditional case as follows:

Definition 3.2 Suppose $\mathbf{a} = (a_1, \ldots, a_i, \ldots, a_n)$ is a fixed action profile. For each X_i, let $\mathbf{a}'_i = (a_1, \ldots, a'_i, \ldots, a_n)$; that is, $\mathbf{a}'_i$ differs from $\mathbf{a}$ in the ith position. The action profile $\mathbf{a}$ is a *conditioned Nash equilibrium* if

$$u_{X_i | \mathrm{pa}(X_i)}(\mathbf{a} | \underbrace{\mathbf{a}, \ldots, \mathbf{a}}_{p_i \text{ times}}) \geq u_{X_i | \mathrm{pa}(X_i)}(\mathbf{a}' | \mathbf{a}, \ldots, \mathbf{a})$$

for all $a'_i \neq a_i$, $i = 1, \ldots, n$. If all utilities are categorical, then the conditioned Nash equilibrium becomes a classical Nash equilibrium. □

A conditioned Nash equilibrium permits each agent to define a rational choice for itself that also takes into consideration the interests of others, and therefore allows the agent to extend its sphere of interest beyond narrow self-interest, but does not lead to a notion of group preference.

3.2 Concurrent decisions

Since the concordant utility is a function of the conjectured action profiles of all players and these profiles may all be different from each other, this utility does not provide a basis for making decisions unless we constrain all of the profiles to be the same. In that case, the concordant utility provides a measure of benefit (the severity of controversy) for the group if all participants were to agree on the same profile.

Definition 3.3 Given a concordant utility $U_{X_1 \cdots X_n}$, a *concurrent group utility* is defined as the concordant utility evaluated at a common profile:

$$c_{X_1 \cdots X_n}(\mathbf{a}) = U_{X_1 \cdots X_n}(\mathbf{a}, \ldots, \mathbf{a}). \tag{3.1}$$

A *concurrence*, denoted $\mathbf{a}_c$, is a profile that maximizes the concurrent utility:

$$\mathbf{a}_g = \arg\max_{\mathbf{a}} c_{X_1 \cdots X_n}(\mathbf{a}). \tag{3.2}$$

$\square$

A concurrence maximizes the functionality of the group in the sense that it represents the single action profile that minimises the severity of controversy, regardless of the benefit to its members. If all members of the group were intent on, and only on, maximizing concordance, they would unanimously choose a concurrence.

3.3 Marginalization

During the process of aggregation, the local social relationships, as characterized by the ex ante conditional utilities, propagate throughout the group to form the concordant utility $U_{\mathcal{X}_n}$. The ex post unconditional utilities u_{X_i}, $i = 1, \ldots, n$ are then extracted from the concordant utility by marginalization according to (2.14), repeated here for convenience:

$$u_{X_i}(\mathbf{a}_i) = \sum_{\sim \mathbf{a}_i} U_{\mathcal{X}_n}(\mathbf{a}_1, \ldots, \mathbf{a}_n). \tag{3.3}$$

The ex post utilities take into consideration all of the social relationships that are expressed ex ante via conditional utilities. They define each stakeholder's personal preferences after having taken into account the interests of all stake-holders who influence it. Such an approach, however, while perhaps providing more psychologically realistic individual utilities, does not get us any closer to a notion of group preference or group rationality.[1] Essentially, once the marginal utilities are defined, the history of their creation ceases to be rel-evant to the application of classical techniques. In fact, such a procedure is nothing more than an application of Friedman's (1961) division of labor. Once defined, marginal utilities are unconditional and indistinguishable in structure from ex ante categorical utilities. Consequently, they may be used according to any classical solution concept.

In the interest of completeness, it is important to consider solution concepts based on marginalization, with Nash equilibrium being the most well-known such concept. Despite its tremendous logical power and utility, however, the Nash equilibrium concept suffers from a number of potentially troubling issues, including (a) Nash equilibria are not unique, and (b) Nash equilibrium solutions are prescriptive but not descriptive (they tell players what to do but not how to do it). Nevertheless, Nash equilibrium solutions are well respected and often serve as the standard that defines acceptable behavior.

3.4 Group and individual welfare

Although the solutions to classical games focus on, and only on, what is good for the players individually, when social relationships exist we may also wish to find solutions that are good for the group as well. Classical game theory, however, drives a wedge between the concept of what is good for individuals and what is good for the group. Shubik (1982) warns against the "anthropomorphic trap" of ascribing judgment and choices to a group: "It may be meaningful, in a given setting, to say that group 'chooses' or 'decides' something. It is rather less likely to be meaningful to say that the group 'wants' or 'prefers' something" (p. 124). This sentiment may be true in the context of individual rationality and categorical utilities, but when the preferences of stakeholders are influenced by the preferences of other stakeholders, it is premature to argue that a notion of group wants or preferences cannot be defined.

Again drawing on the probabilistic analogy, if two random variables are independent, then knowing the value of one of them conveys no information

---

[1] There is an analogous situation in the epistemological context. Computing the marginal proba-bility distributions from the joint distribution destroys information in the sense that, while the marginal distributions of a set of random variables can be derived from the joint distribution, the joint distribution cannot be constructed from the marginal distributions unless the random variables are mutually independent.

about the value the other takes. However, if they are not independent, then knowing the value of one of them does indeed say something about the value the other takes. In other words, some notion of group association exists between the two random variables. By the same reasoning, if two stakeholders are not praxeologically independent, then some notion of group association (sociality) exists between them. If so, then it may indeed be meaningful to define a notion of group preferences that does not obviate notions of individual preferences.

In Section 3.2 we introduced a notion of group preference, but at the expense of considering individual preferences, and in Section 3.3 we were able to extract ex post individual utilities from the concordant utility, but no notion of group preference was considered. In this section we consider both preference notions simultaneously.

Let us first consider preference orderings for the group. To identify such an ordering, we must focus on the concordant utility. As mentioned in Chapter 2, however, the concordant utility does not directly serve as the basis for taking action, since it is a function of multiple profiles (conjectures), and only one profile can actually be implemented. Nevertheless, just as we may extract marginals from the concordant utility for each individual, we may also extract a marginal for the group. To proceed, we observe that since each agent can control only its own actions, what is of interest is the utility of all agents *making conjectures over their own action spaces*.

Definition 3.4   Consider the concordant utility $U_{X_1 \cdots X_n}(\mathbf{a}_1, \dots, \mathbf{a}_n)$. Let a_{ij} denote the jth element of $\mathbf{a}_i$; that is, $\mathbf{a}_i = (a_{i1}, \dots, a_{in})$ is X_i's conjecture profile. Next, form the action profile $(a_{11}, \dots, a_{nn})$ by taking the ith element of each X_i's conjecture profile, $i = 1, \dots, n$. Now let us sum the concordant utility over all elements of each $\mathbf{a}_i$ except the iith elements to form the *group welfare function*[2] $v_{X_1 \cdots X_n}$ for $\{X_1, \dots, X_n\}$, yielding

$$v_{X_1 \cdots X_n}(a_{11}, \dots, a_{nn}) = \sum_{\sim a_{11}} \cdots \sum_{\sim a_{nn}} U_{X_1 \cdots X_n}(\mathbf{a}_1, \dots, \mathbf{a}_n), \qquad (3.4)$$

where $\sum_{\sim a_{ii}}$ means the sum is taken over all a_{ij} except a_{ii}.   □

The group welfare function provides a complete ex post description of the social relationships between the members of a multistakeholder group. Unless its members are praxeologically independent, this utility is not simply an

---

[2] The term "social welfare" is heavily used in welfare economics as a measure of the benefit to a group in a social choice context. Since the current discussion is in the general game-theoretic context, we shall use the term "group welfare" in lieu of "social welfare" when discussing the benefit to the group.

aggregation of individual utilities. It represents a true meso-to-macro transformation of individual conditional utilities (the intermediate level depicted in
Figure 1.1) to a group-level utility (the global level) as the individual conditional preferences propagate through the group, resulting in an emergent notion
of group preference that corresponds to the preference ordering introduced with
Definition 2.1. Such a preference ordering is strictly a mathematical notion and
corresponds to the degrees of concordance that the outcomes provide.

Although the group does not act as a single entity, or superplayer, the group
welfare function nevertheless informs each member of the group regarding
the effect of their joint actions on the society. Each member can extract its
own single-player utility as a function of its own action by computing its own
marginal welfare function.

Definition 3.5 The *individual welfare function* v_{X_i} of X_i is the ith marginal
of $v_{X_1 \cdots X_n}$, that is,

$$v_{X_i}(a_i) = \sum_{\sim a_i} v_{X_1 \cdots X_n}(a_1, \ldots, a_n). \tag{3.5}$$

$\square$

3.5 Negotiation

Once the group and individual welfare functions are defined, they may be
used to generate solution concepts that take into consideration the entire social
environment.

Definition 3.6 The *maximum group welfare* solution is

$$\mathbf{a}^* = \arg\max_{\mathbf{a} \in \mathcal{A}} v_{X_1 \cdots X_n}(\mathbf{a}),$$

and the *maximum individual welfare* solution is

$$a_i^\dagger = \arg\max_{a_i \in \mathcal{A}_i} v_{X_i}(a_i),$$

$\square$

If $a_i^\dagger = a_i^*$ for all $i \in \{1, \ldots, n\}$, the action profile is a *consensus* choice. In
general, however, a consensus will not obtain, and negotiation may be required
to reach a compromise solution.

The existence of group and individual welfare functions provides a rational
basis for meaningful negotiations; namely, that any compromise solution must

at least provide each agent with its security level – that is, the maximum amount of benefit it could receive regardless of the decisions that others might make. The security level for X_i is the maximin utility, defined as

$$s_{X_i} = \max_{a_i} \min_{\sim a_i} u_{X_i}(a_1, \ldots, a_i, \ldots, a_n),$$

where u_{X_i} is the ex post utility given by (3.3).

In addition to individual benefit, we must also consider benefit to the group. Although a security level, per se, for the group cannot be defined in terms of a minimum guaranteed benefit (after all, the group itself does not actually make a choice), a possible rationale for minimum acceptable group benefit is that it should never be less than the smallest benefit to the individuals. This approach is consistent with the principles of justice espoused by Rawls (1971), who argues, essentially, that a society as a whole cannot be better off than its least advantaged member. Accordingly, let us define a security level for the group as $s_{X_1 \cdots X_n} = \min_i \{s_{X_i}\}/n$, where we divide by the number of agents since the utility for the group involves n players.

Now define the *group negotiation set*

$$N_{X_1 \cdots X_n} = \{\mathbf{a} \in \mathcal{A} \colon v_{X_1 \cdots X_n}(\mathbf{a}) \geq s_{X_1 \cdots X_n}\},$$

the *individual negotiation sets*

$$N_{X_i} = \{a_i \in \mathcal{A}_i \colon v_{X_i}(a_i) \geq s_{X_i}\}, \; i = 1, \ldots, n,$$

and the *negotiation rectangle*

$$R_{X_1 \cdots X_n} = N_{X_1} \times \cdots \times N_{X_n}.$$

The negotiation rectangle is the set of profiles such that each member's element provides it with at least its security level. Finally, we define the *compromise set*

$$C_{X_1 \cdots X_n} = N_{X_1 \cdots X_n} \cap R_{X_1 \cdots X_n},$$

which simultaneously provides each member of the group at least its security level, as well as meeting the group's security level. If $C_{X_1 \cdots X_n} = \varnothing$, then no compromise is possible at the stated security levels. One way to overcome this impasse is to decrement the security level of the group iteratively by a small amount, thereby enlarging $N_{X_1 \cdots X_n}$ until $C_{X_1 \cdots X_n} \neq \varnothing$. If $C_{X_1 \cdots X_n} = \varnothing$ after the maximum reduction in group security has been reached, then no compromise is possible, and the group may be considered dysfunctional. Another way to negotiate is for individual members to iteratively decrement their security levels.

Once $C_{X_1\cdots X_n} \neq \varnothing$, any element of this set provides each member, as well as the group, with at least its security level. If $C_{X_1\cdots X_n}$ contains multiple elements, then a tie must be broken. One possible tie-breaker is

$$\mathbf{a}_c = \arg \max_{\mathbf{a} \in C_{X_1\cdots X_n}} v_{X_1\cdots X_n}(\mathbf{a}),$$

which provides the maximum benefit to the group such that each of its members achieves at least its security level.

3.6 Illustrative example

We illustrate the various solution concepts with the Prisoner's Dilemma game introduced in Section 1.2, with the ordinal payoff matrix displayed in Table 1.1. The issue with conventional PD utilities is that X_1's preference for (D, C) is not tempered by the fact that this outcome, while best for itself, is worst for X_2. Similarly, X_2 does not take into account the effect of (C, D) on X_1. Whereas these preferences are entirely consistent with the doctrine of individual rationality, they preclude any rational possibilities of cooperation. These preferences are appropriate if the players are indeed motivated by, and only by, selfish considerations and no social relationship exists between them. Any attempt to reorder the preferences for either player to account for social relationships (such as incorporating Edgeworth's sympathy coefficients) changes the payoffs so that the game is no longer a PD.

As discussed in Section 1.2.3, however, the existence of social influence can modify a players's preferences so they are not completely aligned with its payoffs. Accordingly, we propose an alternative model for the Prisoner's Dilemma that retains the essence of the classical game, but also introduces parameters that modulate the players' preferences. We endow each X_i, $i = 1, 2$, with a *cooperation index* α_i and an *exploitation index* β_i, such that $\alpha_i \in [0, 1]$ measures the degree to which X_i is willing to cooperate, and $\beta_i \in [0, 1]$ is the degree to which X_i is willing to exploit another's vulnerability. To prohibit the contradictory attitudes of simultaneous high cooperation and high exploitation, we restrict $\alpha_i + \beta_i < 1$. To embed this game into a social context, let us assume a hierarchical structure, such that X_1 possesses a categorical utility and X_2 possesses a conditional utility. Although other scenarios are possible, this structure is sufficient to illuminate the possibilities for cooperation.

Let X_1 define its categorical utility as

$$u_{X_1}(C, C) = \alpha_1 \qquad u_{X_1}(C, D) = 0$$
$$u_{X_1}(D, C) = \beta_1 \qquad u_{X_1}(D, D) = 1 - \alpha_1 - \beta_1.$$

Table 3.1. *The conditional utility* $u_{X_2|X_1}(a_{21}, a_{22}|a_{11}, a_{12})$ *for the Prisoner's Dilemma game*

(a_{21}, a_{22})	(a_{11}, a_{12})			
	(C,C)	(C,D)	(D,C)	(D,D)
(C,C)	1	α_2	α_2	0
(C,D)	0	β_2	β_2	0
(D,C)	0	0	0	0
(D,D)	0	$1 - \alpha_2 - \beta_2$	$1 - \alpha_2 - \beta_2$	1

Table 3.2. *The concordant utility* $U_{X_1 X_2}[(a_{11}, a_{12}), (a_{21}, a_{22})]$ *for the Prisoner's Dilemma game*

(a_{11}, a_{12})	(a_{21}, a_{22})			
	(C,C)	(C,D)	(D,C)	(D,D)
(C,C)	α_1	0	0	0
(C,D)	0	0	0	0
(D,C)	$\alpha_2 \beta_1$	$\beta_1 \beta_2$	0	$\beta_1 - \alpha_2 \beta_1 - \beta_1 \beta_2$
(D,D)	0	0	0	$1 - \alpha_1 - \beta_1$

To compute $u_{X_2|X_1}$, we construct four conditional utilities as follows. If X_1 were to conjecture either (C,C) or (D,D) then X_2 should place all of its conditional utility mass on the same joint outcome. If X_1 were to conjecture (C,D), then X_2 should do the same to the degree that it exploits X_1; that is, it should assign β_2 to (C,D), α_2 to (C,C), and $1 - \alpha_2 - \beta_2$ to (D,D). Finally, if X_1 were to conjecture (D,C), that is, to exploit X_2, then X_2 should place zero conditional utility mass on (D,C), β_2 on (C,D), α_2 on (C,C), and $1 - \alpha_2 - \beta_2$ on (D,D). The values of the conditional utility $u_{X_2|X_1}$ are given in Table 3.1. The concordant utility is computed via the aggregation theorem, with values displayed in Table 3.2.

3.6.1 Conditioned Nash equilibria

To compute the conditioned Nash equilibrium, we must examine the four possible outcomes. For outcome (C,C), we see that $u_{X_1}(C,C) > u_{X_1}(D,C)$ if $\alpha_1 >$

β_1, and $u_{X_2|X_1}(C,C|C,C) > u_{X_2|X_1}(C,D|C,C)$; thus, (C,C) is a conditioned Nash equilibrium when $\alpha_1 > \beta_1$. For outcome (C,D), $u_{X_1}(C,D) < u_{X_1}(D,D)$; thus, (C,D) can never be a Nash equilibrium. Also, since $u_{X_2|X_1}(D,C|D,C) < u_{X_2|X_1}(D,D|D,C)$, (D,C) can never be a Nash equilibrium. Finally, since $u_{X_1}(D,D) > u_{X_1}(D,C)$ and $u_{X_2|X_1}(D,D|D,D) > u_{X_2|X_1}(D,C|D,D)$, (D,D) is always a Nash equilibrium.

3.6.2 Concurrence

Although the concordant utility cannot be used directly to render decisions, it does provide a measure of the degree of controversy that exists among the players as they conjecture various outcomes. To illustrate this behavior, let us consider five cases.

$\alpha_1 = 1$. When X_1 is a pure cooperator, concordance is maximized when both players conjecture mutual cooperation, regardless of X_2's propensities.

$\alpha_1 = \beta_1 = 0$. If X_1 is purely passive, that is, has no propensity to either cooperate or exploit, then concordance is maximized with both players conjecture mutual defection, regardless of X_2's propensities.

$\beta_1 = \alpha_2 = 1$. If X_1 is a pure exploiter and X_2 is a pure cooperator, then concordance is maximized by when X_1's conjecture is exploitative and X_2 conjectures mutual cooperation.

$\beta_1 = \beta_2 = 1$. If both players are pure exploiters, then concordance is maximized when the players conjectures are diametrically opposed.

$\beta_1 = 1$, $\alpha_2 = \beta_2 = 0$. If X_1 is purely exploitive and X_2 is purely passive, then concordance is maximized when X_1's conjecture is exploitive and X_2 conjectures mutual defection.

Inspection of Table 3.2 reveals that the utility of a concurrence is as follows:

$$c_{X_1X_2}(C,C) = \alpha_1$$

$$c_{X_1X_2}(C,D) = 0$$

$$c_{X_1X_2}(D,C) = 0$$

$$c_{X_1X_2}(D,D) = 1 - \alpha_1 - \beta_1.$$

We observe that mutual cooperation is a concurrence when $\alpha_1 > 1 - \alpha_1 - \beta_1$, or when $\alpha_1 > \frac{1-\beta_1}{2}$. Similarly, mutual defection is a concurrence when $\alpha_1 < \frac{1-\beta_1}{2}$, and both (C,C) and (D,D) are concordances when $\alpha_1 = \frac{1-\beta_1}{2}$.

Table 3.3. *The* ex post *payoff matrix for the Prisoner's Dilemma game*

	X_2	
X_1	C	D
C	$(\alpha_1, \alpha_1 + \alpha_2\beta_1)$	$(0, \beta_1\beta_2)$
D	$(\beta_1, 0)$	$(1 - \alpha_1 - \beta_1, \ 1 - \alpha_1 - \alpha_2\beta_1 - \beta_1\beta_2)$

3.6.3 Marginalization

Since X_1 possesses a categorical utility, its ex post utility corresponds to its ex ante categorical utility. The ex post marginal utility for X_2, following (3.3), is

$$u_{X_2}(C,C) = \alpha_1 + \alpha_2\beta_1 \qquad u_{X_2}(C,D) = \beta_1\beta_2$$
$$u_{X_2}(D,C) = 0 \qquad u_{X_2}(D,D) = 1 - \alpha_1 - \alpha_2\beta_1 - \beta_1\beta_2.$$

We may now form the ex post payoff matrix by juxtaposing u_{X_1} and u_{X_2}, as illustrated in Table 3.3.

The game defined by these ex post utilities will be a PD if the orderings of the elements correspond to the orderings displayed in Table 1.1. By inspection, the game is a PD if

$$0 < \frac{1 - \beta_1}{2} < \alpha_1 < \beta_1$$
$$0 < 1 - \alpha_1 - \alpha_2\beta_1 - \beta_1\beta_2 < \alpha_1 + \alpha_2\beta_1 < \beta_1\beta_2.$$

For the case where $\alpha_1 = \alpha_2 = \alpha$ and $\beta_1 = \beta_2 = \beta$, the (α, β) region corresponding to the PD game is illustrated in Figure 3.1. This figure also displays the concurrence regions. The region above the line $\alpha = \frac{1-\beta}{2}$ and below the line $\alpha = 1 - \beta$ is the region where mutual cooperation is the concurrence, and the region below the line $\alpha = \frac{1-\beta}{2}$ is the concurrence region for mutual defection.

When the PD inequalities do not hold, we may still examine the game for ex post Nash equilibria. From Table 3.3 we see that (D, D) is an ex post Nash equilibrium for all admissible values of α_i and β_i. We also see that (C, C) is an ex post Nash equilibrium when $\alpha_i > \beta_i$, $i = 1, 2$.

Table 3.4. *Negotiation results for various values of* (α_i, β_i),
$i = 1, 2$ *for the Prisoner's Dilemma game*

$(\alpha_1, \beta_1), (\alpha_2, \beta_2)$	$N_{X_1 X_2}$	$R_{X_1 X_2}$	$C_{X_1 X_2}$
$(0.3, 0.7), (0.3, 0.7)$	$\{(C, C), (D, D)\}$	$\{(D, C)\}$	$\varnothing$
$(0.2, 0.8), (0.2, 0.8)$	$\{(C, C), (D, D)\}$	$\{(D, D)\}$	$\{(D, D)\}$
$(0.3, 0.7), (0.7, 0.3)$	$\{(C, C), (D, C)\}$	$\{(D, C)\}$	$\{(D, C)\}$
$(0.7, 0.3), (0.5, 0.5)$	$\{(C, C), (D, D)\}$	$\{(C, C)\}$	$\{(C, C)\}$

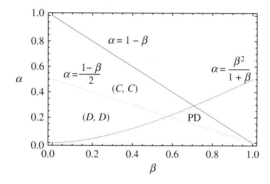

Figure 3.1. Region of α and β for the Prisoner's Dilemma game.

3.6.4 Group and individual welfare solutions

The group welfare function is, from (3.4),

$$v_{X_1 X_2}(C, C) = \alpha_1$$
$$v_{X_1 X_2}(C, D) = 0$$
$$v_{X_1 X_2}(D, C) = \alpha_2 \beta_1$$
$$v_{X_1 X_2}(D, D) = 1 - \alpha_1 - \alpha_2 \beta_1,$$

and the individual welfare functions are, from (3.5),

$$v_{X_1}(C) = \alpha_1 \quad v_{X_1}(D) = 1 - \alpha_1$$
$$v_{X_2}(C) = \alpha_1 + \alpha_2 \beta_1 \quad v_{X_2}(D) = 1 - \alpha_1 - \alpha_2 \beta_1.$$

Table 3.4 illustrates the group negotiation set, the negotiation rectangle, and the compromise set for various values of (α_i, β_i), $i = 1, 2$. For the first case, the values result in an impasse, where the group preference is not reconciled with the individual preferences. For the remaining three cases, a reconciliation is achieved, each with a different compromise state.

3.7 Sociation

The introduction of conditional utilities increases the complexity of the model of a multistakeholder decision problem. To illustrate, consider a three-agent group $\{X_1, X_2, X_3\}$, with utilities u_{X_1}, $u_{X_2|X_1}$, and $u_{X_3|X_1X_2}$, with action spaces $\mathcal{A}_1 = \{a_{11}, a_{12}\}$, $\mathcal{A}_2 = \{a_{21}, a_{22}\}$, and $\mathcal{A}_3 = \{a_{31}, a_{12}\}$. X_1's cardinal utility may assume $2^3 = 8$ possible values. X_2's conditional utility $u_{X_2|X_1}$, however, may assume eight values for each of the eight possible conjectures by X_1, resulting in 64 value specifications. Finally, since X_3's preferences are conditioned on the preferences of both X_1 and X_2, $u_{X_3|X_1X_2}$ requires the specification of eight utility values for each of the 64 possible joint conjectures by X_1 and X_2, thus requiring the specification of 512 values. Clearly, the specification of a total of 584 specifications is not practical. Thus, accounting for conditioning in a multistakeholder decision problem can lead to a combinatorial explosion that quickly renders a problem intractable.

As we see, defining the utilities of a conditional game can quickly get out of hand. If extreme complexity is required to define social relationships properly, the full power of conditional game theory may be necessary. But if not, then it is important to find ways to reduce complexity as much as possible. Thus far, our development has assumed the full generality of conditioning; namely, that a conditional utility depends on the entire conjecture profiles of all of the parents, and the conditional utility is a function of all elements of the action profile. It can be the case, however, that an agent does not condition its preferences on the entire conjecture profiles of its parents. It can also be the case that an agent's utility does not depend upon the entire action profile. To account for such situations, we introduce the notion of sociation.

Suppose X_i has $p_i > 0$ parents $\mathrm{pa}\,(X_i) = \{X_{i_1}, \dots, X_{i_{p_i}}\}$, with conditional utility $u_{X_i|\mathrm{pa}(X_i)}(\mathbf{a}_i|\mathbf{a}_{i_1}, \dots, \mathbf{a}_{i_{p_i}})$, where the joint conjecture $\{\mathbf{a}_{i_1}, \dots, \mathbf{a}_{i_{p_i}}\}$ comprises the conjectures of $\mathrm{pa}\,(X_i)$.

Definition 3.7 A *conjecture subprofile* for X_{i_k}, denoted $\hat{\mathbf{a}}_{i_k}$, is the subset of $\mathbf{a}_{i_k}$ that influences X_i's conditional utility. We then have

$$u_{X_i|\mathrm{pa}(X_i)}(\mathbf{a}_i|\mathbf{a}_{i_1}, \dots, \mathbf{a}_{i_{p_i}}) = u_{X_i|\mathrm{pa}(X_i)}(\mathbf{a}_i|\hat{\mathbf{a}}_{i_1}, \dots, \hat{\mathbf{a}}_{i_{p_i}}).$$

$\{X_1, \dots, X_n\}$ is *completely conjecture sociated* if $\hat{\mathbf{a}}_{i_k} = \mathbf{a}_{i_k}$ for $k = 1, \dots, p_i$ and $i = 1, \dots, n$. It is *completely conjecture dissociated* if $\hat{\mathbf{a}}_{i_k} = a_{i_k}$ for $k = 1, \dots, p_i$ and $i = 1, \dots, n$, in which case,

$$u_{X_i|\mathrm{pa}(X_i)}(\mathbf{a}_i|\mathbf{a}_{i_1}, \dots, \mathbf{a}_{i_{p_i}}) = u_{X_i|\mathrm{pa}(X_i)}(\mathbf{a}_i|a_{i_1}, \dots, a_{i_{p_i}}).$$

Otherwise, the group is *partially conjecture dissociated*. □

Definition 3.8 A *utility subprofile*, denoted $\tilde{\mathbf{a}}_i$, comprises the subset of actions by X_j, $j = 1,\ldots,n$, that affect X_i's utility. We then have

$$u_{X_i \mid \mathrm{pa}(X_i)}(\mathbf{a}_i \mid \mathbf{a}_{i_1}, \ldots, \mathbf{a}_{i_{p_i}}) = \tilde{u}_{X_i \mid \mathrm{pa}(X_i)}(\tilde{\mathbf{a}}_i \mid \hat{\mathbf{a}}_{i_1}, \ldots, \hat{\mathbf{a}}_{i_{p_i}}),$$

where $\tilde{u}$ denotes u with the dissociated elements of its argument removed. $\{X_1, \ldots, X_n\}$ is *completely utility sociated* if $\tilde{\mathbf{a}}_i = \mathbf{a}_i$ for $i = 1, \ldots, n$. It is *completely utility dissociated* if $\tilde{\mathbf{a}}_i = a_i$ for $i = 1, \ldots, n$, in which case

$$u_{X_i \mid \mathrm{pa}(X_i)}(\mathbf{a}_i \mid \mathbf{a}_{i_1}, \ldots, \mathbf{a}_{i_{p_i}}) = \tilde{u}_{X_i \mid \mathrm{pa}(X_i)}(a_i \mid \hat{\mathbf{a}}_{i_1}, \ldots, \hat{\mathbf{a}}_{i_{p_i}}).$$

Otherwise, the group is *partially utility dissociated*. $\square$

For a partially sociated group, the concordant utility assumes the form

$$U_{X_1 \cdots X_n}(\mathbf{a}_1, \ldots, \mathbf{a}_n) = \tilde{U}_{X_1 \cdots X_n}(\tilde{\mathbf{a}}_1, \ldots, \tilde{\mathbf{a}}_n)$$

$$= \prod_{i=1}^{n} \tilde{u}_{X_i \mid \mathrm{pa}(X_i)}(\tilde{\mathbf{a}}_i \mid \hat{\mathbf{a}}_{i_1}, \ldots, \hat{\mathbf{a}}_{i_{p_i}}),$$

where $\tilde{U}$ is U with the dissociated arguments removed.

Definition 3.9 A group $\{X_1, \ldots, X_n\}$ is *completely dissociated* if it is both completely conjecture dissociated and completely utility dissociated, in which case,

$$u_{X_i \mid \mathrm{pa}(X_i)}(\mathbf{a}_i \mid \mathbf{a}_{i_1}, \ldots, \mathbf{a}_{i_{p_i}}) = \tilde{u}_{X_i \mid \mathrm{pa}(X_i)}(a_i \mid a_{i_1}, \ldots, a_{i_{p_i}}).$$

$\square$

For a completely dissociated group, the concordant utility coincides with the group welfare function and assumes the form

$$v_{X_1 \cdots X_n}(a_1, \ldots, a_n) = \prod_{i=1}^{n} \tilde{u}_{X_i \mid \mathrm{pa}(X_i)}(a_i \mid a_{i_1}, \ldots, a_{i_{p_i}}). \tag{3.6}$$

Example 3.1 Let us consider a three-agent case where X_1 possesses a categorical utility, X_2's utility is conditioned on X_1, and X_3's utility is conditioned on both X_1 and X_2. First, we consider the completely sociated case; namely, that the conditional utility for X_2 is a function of its entire action profile and of the entire conjecture of X_1, and that the conditional utility for X_3 is a function of its entire profile and of the entire joint conjecture of X_1 and X_2. The corresponding concordant utility is then of the form

$$U_{X_1 X_2 X_3}[(a_{11},a_{12},a_{13}),(a_{21},a_{22},a_{23}),(a_{31},a_{32},a_{33})] =$$

$$u_{X_1}(a_{11},a_{12},a_{13})u_{X_2|X_1}(a_{21},a_{22},a_{23}|a_{11},a_{12},a_{13})$$

$$u_{X_3|X_1 X_2}[a_{31},a_{32},a_{33}|(a_{11},a_{12},a_{13}),(a_{21},a_{22},a_{23})].$$

Second, let us assume a conjecture dissociated case, such that X_2's utility is conditioned only on X_1's conjecture of a_{11}, and X_3's utility is conditioned only on X_1's conjecture of a_{11} and on X_2's conjecture on a_{22}, in which case the concordant utility is

$$U_{X_1 X_2 X_3}[(a_{11},a_{12},a_{13}),(a_{21},a_{22},a_{23}),(a_{31},a_{32},a_{33})] =$$

$$u_{X_1}(a_{11},a_{12},a_{13})u_{X_2|X_1}(a_{21},a_{22},a_{23}|a_{11})$$

$$u_{X_3|X_1 X_2}(a_{31},a_{32},a_{33}|a_{11},a_{22}).$$

Third, let us assume, in addition to partial conjecture dissociation, that X_1's utility is a function of a_{11} and a_{12}; that is, X_1 is partially utility dissociated. We also assume that X_2's utility is conditioned on X_1's conjecture regarding both a_{11} and a_{12} but, given that conditioning, X_2's utility is a function only of a_{22}; that is, X_2 is completely utility dissociated. We also assume that X_3's conditional utility is a function only of a_{31} and a_{33}, conditioned on a_{11} and a_{22}. The resulting concordant utility is then

$$\tilde{U}_{X_1 X_2 X_3}(a_{11},a_{12},a_{22},a_{31},a_{33}) = \tilde{u}_{X_1}(a_{11},a_{12})\tilde{u}_{X_2|X_1}(a_{22}|a_{11},a_{12})$$

$$\tilde{u}_{X_3|X_1 X_2}(a_{31},a_{33}|a_{11},a_{22}),$$

from which the group welfare function can be obtained as

$$v_{X_1 X_2 X_3}(a_{11},a_{22},a_{33}) = \sum_{a_{12},a_{31}} \tilde{U}_{X_1 X_2 X_3}(a_{11},a_{12},a_{22},a_{31},a_{33}).$$

Finally, consider the completely dissociated case, where X_1's utility is a function only of a_{11}, X_2's utility is a function only of a_{22} given X_1's conjecture on a_{11}, and X_3's utility is a function only of a_{33} given X_1's conjecture on a_{11} and X_2's conjecture on a_{22}. With the completely dissociated case, the concordant utility coincides with the group welfare function, yielding

$$v_{X_1 X_2 X_3}(a_{11},a_{22},a_{33}) = \tilde{u}_{X_1}(a_{11})\tilde{u}_{X_2|X_1}(a_{22}|a_{11})\tilde{u}_{X_3|X_1 X_2}(a_{33}|a_{11},a_{22}).$$

□

3.8 Nondominance

When more than a single stakeholder is involved in a decision problem, the notion of optimality is not uniquely defined. Barring the unlikely circumstance that there exists an alternative that simultaneously maximizes the benefit to all stakeholders, there will exist a set of extremizing alternatives, each of which is optimal from the point of view of some stakeholder. Historically, the first to express this concept analytically appears to have been Edgeworth (1881): "It is required to find a point $(x\ y)$ such that, *in whatever direction* we take an infinitely small step, P and Π do not increase together, but that, while one increases, the other decreases [emphasis in original]" (p. 21). Credit for this concept, however, is more commonly given to Pareto, who stated in 1927 what is commonly known as the principle of Pareto (1927/1971) optimality.

> We will say that the members of a collectivity enjoy *maximum ophelimity* in a certain portion when it is impossible to find a way of moving from that position very slightly in such a manner that the ophelimity enjoyed by each of the individuals of the collectivity increases or decreases. That is to say, any small displacement in departing from that position necessarily has the effect of increasing the ophelimity which certain individuals enjoy, of being agreeable to some and disagreeable to others [emphasis in original]. (p. 261)

A decision-making procedure common in the operations research context is the concept of nondomination, which leads to the so-called Pareto frontier. We first introduce this concept in the classical context of categorical utilities and then extend it to the conditional context.

Definition 3.10 Let $\mathcal{X}_n = \{X_1, \ldots, X_n\}$ denote a group such that categorical utilities of the form u_{X_i} exist for $i = 1, \ldots, n$. An action profile **a** *dominates* the action profile **a**$'$ if $u_{X_i}(\mathbf{a}) \geq u_{X_i}(\mathbf{a}')$ for all i, with strict inequality holding for at least one $i \in \{1, \ldots, n\}$. A profile **a** *strongly dominates* **a**$'$ if $u_{X_i}(\mathbf{a}) > u_{X_i}(\mathbf{a}')$ for all $i \in \{1, \ldots, n\}$. The profile **a** is *nondominated* if it is not dominated by any other profile. The *Pareto frontier*[3] is the set of all nondominated action profiles. A profile is said to be *Pareto efficient* if it is member of the Pareto frontier. □

Thus, one action profile dominates another if it is no worse for all stakeholders, and strictly better for at least one stakeholder, and strongly dominates another if it is strictly better for all stakeholders.

The notion of dominance generates a partial ordering $\succeq_D$ between the elements of $\mathcal{A}$; that is, $\mathbf{a} \succeq_D \mathbf{a}'$ if **a** dominates **a**$'$. This ordering is *not reflexive*, since it cannot happen that $\mathbf{a} \succeq_D \mathbf{a}$ (there must be at least one strict inequality).

---

[3] Although Edgeworth's (1881) research predates Pareto's (1927) publication, history has seen fit to attach the latter's name to this concept – a historical accident that is not easily changed. However, it is important at least to give Edgeworth proper credit for his discovery.

The ordering is, however, *asymmetric*, since if $\mathbf{a} \succeq_D \mathbf{a}'$, then $\mathbf{a}' \not\succeq_D \mathbf{a}$. Furthermore, the ordering is transitive, since if $\mathbf{a} \succeq_D \mathbf{a}'$ and $\mathbf{a}' \succeq_D \mathbf{a}''$, then $\mathbf{a} \succeq_D \mathbf{a}''$. Thus, $\succeq_D$ is a strict partial order. It is *not* a strict order, however, since it is not complete: $\mathbf{a} \not\succeq_D \mathbf{a}'$ does not imply $\mathbf{a}' \succeq_D \mathbf{a}$. Finally, since $\succeq_D$ is not reflexive, it is also not a partial order.

Computationally, identifying the Pareto frontier is a challenging search problem. The naive, globally exhaustive procedure of computing all possible dominance ordering tests, each comprising n comparisons, requires on the order of $n|\mathcal{A}|^2$ computations, where $|\mathcal{A}|$ is the cardinality of the product action space. Efficient approaches have been devised, however, that take on the order of $\mathcal{A}(\log |\mathcal{A}|)^{n-2}$ computations for $n \geq 4$ and $|\mathcal{A}|(\log |\mathcal{A}|)$ computations for $n = 2, 3$ (Kung et al., 1975).[4]

Elements of the Pareto frontier possess a weak notion of optimality in the sense that if any one individual attempts to change its action in an attempt to improve its payoff, then the payoff of at least one other individual will decrease. Essentially, restricting attention to the Pareto frontier eliminates all actions for which an obviously better one exists. Pareto efficiency, therefore, is often viewed as a minimal condition for any action profile to be seriously considered as an acceptable outcome.

Pareto efficiency, however, does not imply that any member of the Pareto frontier should be viewed as a measure of benefit to the group as a whole – it is not a measure of group welfare. Nevertheless, the Pareto frontier provides valuable information that may be used, for example, by a tie-breaking procedure, to choose a profile that admits an exogenous concept of benefit to the group.

When extending the notion of Pareto efficiency to the context of conditional utilities, we immediately face a problem. Given a conditional utility $u_{X_i | \text{pa}(X_i)}(\mathbf{a}_i | \mathbf{a}_{i_1}, \ldots, \mathbf{a}_{i_{p_i}})$, where the joint conjecture $\{\mathbf{a}_{i_1}, \ldots, \mathbf{a}_{i_{p_i}}\}$ corresponds to the p_i parents $\text{pa}(x_i) = \{X_{i_1}, \ldots, X_{i_{p_i}}\}$ of X_i, we must extend the notion of dominance to include dominance with respect to all joint conjectures of the parents of X_i. Such a general concept is of limited value, and is likely to be computationally as well as conceptually complex. Under the condition of complete conjecture dissociation, however, it is possible to define a useful notion of conditional Pareto efficiency.

---

[4] The complexity of searching for the Pareto frontier when the dimensionality is large has motivated extensive research into ways to address this problem. The classical approach is to focus on ways to transform a multistakeholder optimization problem into a single-objective optimization problem where conventional optimization techniques can be applied. Another approach, called the evolutionary method, replaces the direct search for optimal solutions with an iterative random searching technique is designed to converge to the Pareto frontier.

Table 3.5. *Environmental impact utility* $u_{X_1}(P_i)$

P_1	P_2	P_3	P_4
0.3571	0.2857	0.2143	0.1429

Definition 3.11 Let $\{X_1,\ldots,X_n\}$ be a completely conjecture dissociated group, and let $\mathrm{pa}\,(X_i) = \{X_{i_1},\ldots,X_{i_{p_i}}\}$ denote the p_i parents of X_i. For a conjecture $\mathbf{a} = (a_1,\ldots,a_n)$, let $(a_{i_1},\ldots,a_{i_{p_i}})$ denote the subprofile corresponding to $\mathrm{pa}\,(X_i)$. The profile $\mathbf{a}$ *conditionally dominates* $\mathbf{a}' = (a'_1,\ldots,a'_n)$ if

$$u_{X_i|\mathrm{pa}\,(X_i)}(\mathbf{a}|a_{i_1},\ldots,a_{i_{p_i}}) \geq u_{X_i|\mathrm{pa}\,(X_i)}(\mathbf{a}'|a'_{i_1},\ldots,a'_{i_{p_i}})$$

with strict inequality for at least one $i \in \{1,\ldots,n\}$. A profile $\mathbf{a}$ *strongly conditionally dominates* $\mathbf{a}'$ if

$$u_{X_i|\mathrm{pa}\,(X_i)}(\mathbf{a}|a_{i_1},\ldots,a_{i_{p_i}}) > u_{X_i|\mathrm{pa}\,(X_i)}(\mathbf{a}'|a'_{i_1},\ldots,a'_{i_{p_i}})$$

for all $i \in \{1,\ldots,n\}$. The *conditioned Pareto frontier* is the set of all conditionally nondominated action profiles. A profile is said to be *conditionally Pareto efficient* if it is a member of the conditioned Pareto frontier. □

An important special case obtains when all participants are completely utility dissociated as well, in which case the action profile $\mathbf{a} = (a_1,\ldots,a_n)$ conditionally dominates $\mathbf{a}' = (a'_1,\ldots,a'_n)$ given $\{a_{i_1},\ldots,a_{i_{p_i}}\}$ if $\tilde{u}_{X_i|\mathrm{pa}\,(X_i)}(a_i|a_{i_1},\ldots,a_{i_{p_i}}) \geq \tilde{u}_{X_i|\mathrm{pa}\,(X_i)}(a'_i|a_{i_1},\ldots,a_{i_{p_i}})$.

Example 3.2 A factory can produce one of four different widgets, $\{W_1, W_2, W_3, W_4\}$, each of which can be manufactured by any of four different process, $\{P_1, P_2, P_3, P_4\}$. The problem is to choose from among the 16 possible alternatives (P_i, W_j), where $i, j \in \{1,2,3,4\}$. X_1 is responsible for the first phase of the manufacturing process and is deeply concerned with the environment. Each manufacturing process entails an environmental impact, as indicated by the utility values illustrated in Table 3.5, where the higher the utility, the more favorable (less damaging) the impact is on the environment.

X_2 is responsible for the second phase of the process and is primarily concerned with getting the final product to the market. Although each widget may be manufactured by any of the processes, they do not all generate the same profit. Consequently, there is a different profit utility for each process. Table 3.6 displays these utilities, with each column representing the profit utility associated with the corresponding manufacturing process.

Table 3.6. Profit utility $u_{X_2|X_1}(W_j|P_i)$
for each manufacturing process

	P_1	P_2	P_3	P_4
W_1	0.1667	0.0870	0.1852	0.0909
W_2	0.1666	0.3043	0.3333	0.2727
W_3	0.2500	0.4348	0.4444	0.1818
W_4	0.4167	0.1739	0.0370	0.4545

Table 3.7. The group welfare function
$v_{X_1 X_2}(P_i, W_j)$ for the manufacturing
problem

	X_2			
	P_1	P_2	P_3	P_4
W_1	0.0595	0.0249	0.0397	0.0130
W_2	0.0595	0.0869	0.0714	0.0390
W_3	0.0893	0.1241	0.0952	0.0260
W_4	0.1488	0.0497	0.0079	0.0649

Since this group is completely dissociated, the aggregation theorem yields
the group welfare function

$$v_{X_1 X_2}(P_i, W_j) = \tilde{u}_{X_1}(P_i)\tilde{u}_{X_2|X_1}(W_j|P_i),$$

which is displayed in Table 3.7. Notice that this function is also the concordant
utility.

We may also obtain the individual welfare functions by computing the
marginals of the group welfare function. Since X_1 possesses a categorical utility, its ex post marginal utility is the same as its ex ante categorical utility, as
illustrated in Table 3.5. Table 3.8 displays X_2's ex post marginal utility. Since
the utilities for this problem are completely dissociated, the security level for
each X_i is its maximum utility, resulting in a negotiation rectangle of the form
$R_{X_1 X_2} = \{P_1, W_3\}$.

By inspection, the group security level is 0.1673, resulting in a group
negotiation set $N_{X_1 X_2} = \varnothing$. To reach a solution, a compromise must be negotiated. Let us require each individual to lower its security level to include
both the best and next-best choices, resulting in a new negotiation rectangle
$R_{X_1 X_2} = \{(P_1, W_3), (P_1, W_4), (P_2, W_3), (P_2, W_4)\}$. The new security level for the

Table 3.8. *Individual welfare*
function $v_{x_2}(W_j)$ *for the*
manufacturing problem

W_1	W_2	W_3	W_4
0.1371	0.2568	0.3346	0.2713

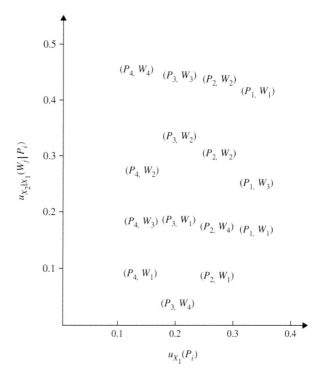

Figure 3.2. The utility space illustrating the conditioned Pareto frontier.

group becomes 0.1356, yielding $N_{X_1 X_2} = \{(P_1, W_4)\}$. Thus, the compromise set is $C_{X_1 X_2} = \{(P_1, W_4)\}$ and $\mathbf{a}_c = (P_1, W_4)$, which corresponds to the best outcome for the group, the best outcome for X_1, and the second-best outcome for X_2. This compromise corresponds to each member of the group being willing to reduce its profit in order to achieve an acceptable environmental impact.

Figure 3.2 displays the utility space constructed by plotting $u_{x_1}(a_1)$ along the abscissa and $u_{x_2 | x_1}(a_2 | a_1)$ along the ordinate. By inspection, the conditioned

Pareto frontier is

$$\{(P_1, W_4), (P_2, W_3), (P_3, W_3), (P_4, W_4)\}.$$

Notice that the compromise solution $\mathbf{a}_c = (P_1, W_4)$ is a member of this set.

$\square$

3.9 Conditional social choice

Our treatment thus far has focused on the general game-theoretic context, where each agent is able to implement its own choice; that is, we deal with action profiles $\mathbf{a} = (a_1, \ldots, a_n) \in \mathcal{A} = \mathcal{A}_1 \times \cdots \times \mathcal{A}_n$ where X_i is empowered to take action a_i, $i = 1, \ldots, n$. In this section we specialize conditional game theory to scenarios where there is only one action set $\mathcal{A}$, from which the group implements a single action. Such a scenario is, by definition, a social choice problem. Broadly speaking, classical social choice theory is a theory of how individual interests can be used to define rational behavior for the group. In this treatment, however, we do not explore the many ways that the social choice problem can be framed. Instead, we focus on only one framing: the case where stakeholders possess cardinal utilities that permit interpersonal comparisons of utility. Within this limited realm we extend the theory to account for conditional preferences, and leave untouched the issue of how to deal with conditioning for the more general framings of social choice theory.

Once we limit attention to this special case, conditional social choice can be viewed as a conditional game with the constraint that there is only one alternative set, denoted $\mathcal{A}$, available to the group.[5] Stakeholders are empowered to express their preferences over $\mathcal{A}$, but these individual preferences must be combined according to some logical rule to identify which alternative will be instantiated by the group.

Definition 3.12 A *social choice conjecture* for X_i (compare with Definition 2.2) is a single alternative $a_i \in \mathcal{A}$, and a *joint social conjecture* is a vector $(a_1, \ldots, a_n) \in \mathcal{A} \times \cdots \times \mathcal{A}$. The *social choice concordant utility* (compare with (2.17)) is

$$W_{X_1 \cdots X_n}(a_1, \ldots, a_n) = \prod_{i=1}^{n} u_{X_i | \mathrm{pa}(X_i)}(a_i | a_{i_1}, \ldots, a_{i_{p_i}}). \tag{3.7}$$

The expression

$$W_{X_1 \cdots X_n}(a_1, \ldots, a_n) > W_{X_1 \cdots X_n}(a_1', \ldots, a_n')$$

----

[5] In the social choice context, we shall refer to the choice set as alternatives.

means that if each X_i were to conjecture the corresponding a_i, the degree of concordance would be greater (preference conflicts would be less severe) than if each X_i were to conjecture a'_i. □

Definition 3.13 A *conditional social choice problem* is a triple $\{\mathcal{X}_n, \mathcal{A}, W_{\mathcal{X}_n}\}$ where $\mathcal{X}_n = \{X_1, \ldots, X_n\}$ is a group of n stakeholders with group-level action space $\mathcal{A}$ and $W_{\mathcal{X}_n}$ is a social choice concordant utility. Equivalently, a conditional social choice problem can be defined in terms of the completely dissociated conditional utilities $u_{X_i | \mathrm{pa}(X_i)}$, $i = 1, \ldots, n$. If all utilities are categorical, a conditional social choice problem becomes a classical social choice problem. □

Solving a conditional social choice problem requires us to define notions of both group and individual welfare.

Definition 3.14 Given a social choice concordant utility $W_{X_1 \cdots X_n}$, the *social welfare function* is defined as the concordant utility evaluated at a common alternative, yielding

$$w_{X_1 \cdots X_n}(a) = W_{X_1 \cdots X_n}(a, \ldots, a). \tag{3.8}$$

The *individual welfare function* for X_i is the ith marginal of the concordant utility; that is,

$$w_{X_i}(a_i) = \sum_{\sim a_i} W_{X_1 \cdots X_n}(a_1, \ldots, a_n). \tag{3.9}$$

□

Once social and individual welfare functions are available, it may be possible to define compromise solutions that are simultaneously acceptable to all stakeholders. As introduced in Section 3.5, we may define security levels for each stakeholder as well as a security level for the group. We may then define the group negotiation set

$$N_{X_1 \cdots X_n} = \{a \in \mathcal{A}: w_{X_1 \cdots X_n}(a) \geq s_{X_1 \cdots X_n}\},$$

and the *individual negotiation sets*

$$N_{X_i} = \{a_i \in \mathcal{A}_i: w_{X_i}(a_i) \geq s_{X_i}\}, \ i = 1, \ldots, n.$$

The *compromise set* is then

$$C_{X_1 \cdots X_n} = N_{X_1 \cdots X_n} \cap N_{X_1} \cdots \cap N_{X_n}.$$

As the game-theoretic context, if $C_{X_1 \cdots X_n} \neq \varnothing$, then a rational compromise is possible, but if $C_{X_1 \cdots X_n} = \varnothing$, then a compromise may be obtained by iteratively decrementing the security levels.

3.10 Summary

Conditional game theory removes, or at least mitigates, the wedge that exists between considerations of group benefit and individual benefit. Classical game theory does not easily accommodate both notions simultaneously, and game theorists are reluctant to tolerate a serious notion of group rationality, for to do so would require acknowledging the existence of a superplayer who can make decisions for the group. Such a notion is fundamentally inconsistent with the concept of autonomous players, each of whom is capable of making choices according to its own concept of rationality. Furthermore, virtually all conventional solution concepts are based on the notion of individual rationality, a concept that perfectly corresponds to the categorical utility structure of classical games.

With the introduction of conditional utilities and conditional games, however, the interests of the players may extend beyond the self, and it becomes possible for a notion of group preference to emerge as the conditional preferences propagate through the group. Using this structure, it is possible to extract a group welfare function that defines the emergent preferences of the group, and at the same time to extract individual welfare functions that define the ex post preferences for the individuals as a result of their interaction.

4

Coordination

A mathematical formalism may be operated in ever new, uncovenanted ways, and force on our hesitant minds the expression of a novel conception.

— Michael Polanyi
Personal Knowledge (University of Chicago Press, 1958)

4.1 Extrinsic versus intrinsic coordination

We begin our discussion in this chapter with a review of the various notions of coordination that appear in the literature. Perhaps the the earliest game-theoretic concept of coordination is that of tacit coordination, as introduced by Schelling (1960). According to this concept, players coordinate if they try to guess what each other will do and make their decisions accordingly. Lewis has provided more structure to this notion by introducing the notion of coordination equilibria. Recall that a Nash equilibrium is an action profile such that no single player would increase its payoff if it acted otherwise alone. Lewis (1969) has refined this idea to define a *coordination equilibrium* as an action profile such that no single player would increase its payoff had *any one* player alone acted otherwise, either itself or someone else. A coordination equilibrium is *proper* if each agent strictly prefers it to any other outcome. Lewis (1969) provides a formal definition of a *coordination problem* in the context of classical game theory: "Coordination problems ... are situations of interdependent decision by two or more agents in which coincidence of interest predominates and in which there are two or more proper coordination equilibria" (p. 24).

Although the notion of coordination equilibria is a refinement of Nash equilibria, such equilibria are not confined to coordination games. In fact, even a zero-sum game can have a coordination equilibrium, as illustrated by the game

Table 4.1. *The payoff matrix for*
a zero-sum game with a
coordination equilibrium

	X_2	
X_1	a_{21}	a_{22}
a_{11}	(0,0)	(0,0)
a_{12}	(0,0)	(1,−1)

whose payoff matrix is displayed in Table 4.1, where, by inspection, we see that the profile (a_{11}, a_{21}) is a coordination equilibrium. Thus, a coordination equilibrium, although perhaps informative, is not a definitive manifestation of coincidence of interest.

Bicchieri (1993) also treats coordination from the perspective of classical game theory but argues that rational choice theory alone is not enough to effect meaningful coordination. It must be augmented with a theory of belief formation – that is, a theory of learning. If players are permitted to play the same game many times, they may gain insight regarding the social dispositions of the other players and may be able to predict their behavior, establish their own reputations, and thereby gain the trust of others to their mutual advantage. As players repeatedly interact in a society, they may learn to recognize behavioral patterns and settle on stable ones. This account of coordination is based on the assumption that players will eventually conform to behavior that is acceptable to all members of the community. Coordination, therefore, is viewed as the end result of social evolution. At the end of the day, however, this notion of coordination relies upon extra-game-theoretic notions of social behavior that are not explicitly present in the mathematical structure of the game. Although this notion of coordination may be a useful analysis model with which to explain human behavior, it is less clear that it is a useful synthesis model for the design of artificially intelligent decision-making agents.

Another systematic way to investigate coordination has been developed by Malone and Crowston (1991, 2003), who provide a taxonomy of coordination consisting of four elements: goals, activities, actors, and interdependencies. They offer the following operational definition: "Coordination is managing interdependencies among activities" (Malone and Crowston, 2003, p. 50). Interdependencies can exist at multiple levels. Perhaps the most basic form of interdependence occurs when the outcome to each individual depends on, and only on, the choices of all individuals, as is the case with games that

Table 4.2. *The payoff matrix for the Heads or Tails game*

X_1	X_2	
	H	T
H	(1,1)	(0, 0)
T	(0, 0)	(1,1)

employ categorical utilities. This definition refers to interdependence in terms of actions, but there is also a more profound notion of interdependence, namely, *the interdependence of preferences.*

We illustrate the difference between these two notions of interdependence by considering the game of Heads or Tails, whose payoff matrix is displayed in Table 4.2. A coin is flipped and the players are each asked to guess the face that appears, either heads, H, or tails, T. This game is a coordination problem as defined by Lewis (1969) and has two proper coordination equilibria: (H, H) and (T, T). We may assess the *action interdependence* as follows. Assuming that both players operate under the assumption that the coin is fair, each of four possible choice profiles – (H, H), (H, T), (T, H), and (T, T) – would be realized with equal probability. Consequently, it would be reasonable to assume that the players would both choose, say, H, with probability 0.25.

Experiments with human respondents, however, reveal that they are more likely to choose H than T. For example, Schelling (1960) reports the results of experiments that indicate that 86% of the respondents guessed H, and he gave the name *focal point* to the equilibrium that is the most likely to be chosen by both players. According to these data, both participants would choose H with probability $0.86 \times 0.86 \approx 0.74$. It is clear from this and other similar studies that the outcomes of such games as this can be heavily influenced by the psychological dispositions of the players.

One way to view this situation is to recognize that although the payoff for both players choosing H is the same as for both choosing T, their preferences are not governed solely by probabilities, as would be the case if action interdependence were the only factor involving coordination (in which case the experimental results would be closer to both choosing H with probability 0.25). In other words, the players have a strong *preference interdependence* that is not motivated by the probabilities or the payoffs.

Table 4.3. *The payoff matrix
of a game with strategic
complements*

	X_2	
X_1	a_{21}	a_{22}
a_{11}	(1,1)	(1,0)
a_{12}	(0,1)	(2,2)

Considerable research has also been conducted on the concept of *strategic complementarity*, as defined by Cooper (1999): "Put in simple words, this condition implies that higher actions by other players provide an incentive for the remaining player to take higher action as well" (p. 19). Decisions are strategic complements if switching to a choice with a higher potential payoff by one player would open up the opportunity for other players to increase their payoffs as well. Consider the game with the payoff matrix illustrated in Table 4.3. This game possesses two coordination equilibria: (a_{11}, a_{21}) and (a_{12}, a_{22}). If X_1 were to switch from a_{11} to a_{12}, it would gain the opportunity to increase its payoff from one unit to two units, but it would also risk reducing its guaranteed payoff from one to zero. However, the possibility of such a switch would be attractive to X_2, thus providing an incentive for it to make the switch. Games that possess strategic complementarity also possess an enhanced incentive for the players to cooperate, but whether this incentive is realized depends heavily on the dispositions of the players. A risk-averse player would be unlikely to respond to such an incentive. Clearly, playing according to strategic complementarity also depends heavily on the social and psychological attributes of the players.

Coordination has also been the focus of research by Goyal (2007), who distinguishes between the notions of coordination and cooperation with the following operational definitions: "The problem of coordination arises in its simplest form when, for an individual, the optimal course of action is to conform to what others are doing ... the problem of cooperation arises when individual incentives lead to an outcome which is socially inefficient or undesirable" (pp. 63, 64). These definitions, however, act to make coordination and cooperation mutually exclusive. Coordination, according to this view, is virtually synonymous with the notion of strategic complementarity: the interests of all players reinforce each other, thus providing incentives to work in harmony. According to this concept of cooperation, however, for a player to cooperate (rather than coordinate), the player's individual interests must be at odds with the interests of others. This is a fine line to draw between cooperation and coordination.

Perhaps the problem with operational definitions of these concepts is that they depend on the sociological orientations of the players. For players who are disposed to work in harmony, there may be little operational difference between cooperation and coordination, but for players who are not so disposed, the difference may be dramatic. Consider, for example, athletic contests or military engagements. Players on opposite teams will certainly try to outguess each other and will rely on imaginative processes of introspection; in other words, they will operate according to Schelling's (1960) notion of tacit coordination. Yet, this aspect of social relationships seems to be excluded under the operational definitions proposed by Goyal (2007).

The concepts of tacit coordination, coordination equilibria, action interdependence, and strategic complementarity are all based on the rationality concepts employed by the players. But rationality, per se, is not part of the mathematical structure of the game. The game is defined by, and only by, the players, their feasible actions, and the payoffs. Notions of rationality, which are employed to define solution concepts, are external to the mathematical game description. Consequently, the above concepts of coordination are *extrinsic*; they are not part of the mathematical structure that defines the game.

We now ask the question: Is it possible to define a concept of coordination that is a function of, and only of, the mathematical structure of a game as distinct from notions of rational behavior? At first glance, it may appear that this question is meaningless. Once the payoffs of all players are juxtaposed to form a payoff array, opportunities for cooperation, compromise, exploitation, competition, benevolence, malevolence, and so forth, are exposed. Since these are all social attributes, the argument goes, they cannot be separated from rationality. Thus, to talk of coordination without appealing to notions of rational behavior is impossible.

That argument, however, depends upon a critical assumption; namely, that the utilities of all players are categorical. If we abandon that condition and permit utilities to be conditional, then it is possible to form social bonds and to create a meaningful notion of group preference and, hence, of group-level rational behavior. Once such a notion can be defined, it becomes possible to apply the dictionary definition of coordination, as introduced in Section 1.3, which asserts that coordination requires the parts (the individual-level preference orderings) to be combined to form a whole (the group-level preference ordering). This possibility leads to a new concept of coordination, one that focuses on, and only on, the mathematical structure of the game. We shall refer to this type of coordination as *intrinsic* coordination.

The key element of intrinsic coordination is the presence of both group and individual preference orderings. Recall, as Arrow's (1951) impossibility theorem attests, that a group preference ordering does not follow as a logical consequence of individual categorical preference orderings. Thus, intrinsic coordination is not possible when all utilities are categorical. By permitting conditional utilities, however, the conditions of Arrow's theorem are no longer satisfied, and it does indeed become possible to define a group preference ordering as the logical consequence of individual conditional preference orderings, as established in Chapter 3.

Intrinsic coordination depends on the social influence that exists among the members of the group and has nothing to do, specifically, with the coincidence, or lack thereof, of the preferences or with the notions of rationality of the stakeholders. Although intrinsic coordination may give rise to cooperative behavior as perceived by an outside observer (or even by the agents themselves), the presence of such behavior itself is not evidence of an intrinsic capacity to coordinate.

Since intrinsic coordination is a function of the utilities only, it is desirable to analyze their structure to determine just how able the group is to arrive at a solution that simultaneously serves the interests both the individuals and the group. To this end, we seek a quantitative measure of the intrinsic ability, or capacity, of a group to coordinate. Such a measure should not depend upon the rationale for making decisions, or upon the skill or the introspective powers of the players to anticipate the behavior of others by sociological or psychological means other than those that can be extracted directly form the payoffs.

To define such a measure, we must investigate the structural properties of the utilities at much greater depth than simply considering the opportunities for cooperation or competition that become apparent once the payoffs of the players are juxtaposed in a payoff array. To begin our search, we first develop a quantitative measure of the difficulty, or *hardness*, associated with a decision. Using that measure, we proceed to define a notion of mutuality among the stakeholders, and use that notion to define a measure of coordination capacity as a function of, and only of, the structure of the utilities used to define the game.

4.2 Hardness

There are many reasons why decision making can be difficult. First, the decision makers may be indecisive or may not be able to define their preferences completely. Second, there may be uncertain components that can only be described probabilistically. Third, the problem may be complex and not completely understood. Fourth, there maybe unresolved conflicts. But let us assume that all stakeholders are able to define their preferences, that all uncertainty has been eliminated, that all complexities are well understood and properly modeled, and that all conflicts have been resolved. Let us even assume that the best choice is clearly identified. Under these ideal circumstances, is the decision problem easy? In one sense, it is, because a rational decision maker is obligated to make the best choice. If the best choice is overwhelmingly best, then the solution is incontrovertible. But what if other alternatives are nearly as good? What if all choices are nearly equally poor, and the decision maker is in a position of making the best of a bad situation? Such questions may remain after all modeling and conceptual issues have been completely resolved, and highlight the intrinsic difficulty of making a decision. They are manifestations of *opportunity cost*, which is defined as the utility of the next-best alternative. Harsanyi (1977) expressed the issue this way:

> Fundamentally the need for choosing among alternative ends arises because in most cases we cannot obtain *all* our ends at the same time: If we choose to pursue one particular end, then we have to forego some others. The loss of some other ends is the *opportunity cost* of choosing this particular end. Our choice of any given end will be rational if it is based on clear awareness and careful evaluation of the opportunity costs involved [emphasis in original]. (p. 8)

It can be argued that an appreciation of the opportunity costs may be nice, but it is essentially irrelevant. Under the paradigm of classical rational choice theory, it is assumed that all of the economic, social, and psychological aspects of the problem will have been carefully considered and incorporated into the decision makers' utilities. Since rational decision makers would bring all available knowledge to bear on the problem before declaring it to be fully defined, there would be no reason to raise questions or demur at the moment of truth when a decision must be made. The problem is what it is, and the decision makers must take whatever comfort is afforded them in the conviction that they are making the best possible choice. Nevertheless, a human decision maker may feel some tension, or even anxiety, at the moment of truth. A decision made with high opportunity cost can be less satisfying than one with small costs. People often use the terms '*difficult*' and '*hard*' to describe such circumstances, and such disquiet may prompt them to explore ways to modify the decision problem, such as demanding additional data, expanding the set of feasible actions, or seeking advice, assistance, or therapy.

As artificially intelligent decision-making entities become increasingly sophisticated, the issue of how to deal with high opportunity costs and other disquieting issues will become more important. Unlike human decision makers, who can define "difficult" and "hard" subjectively, an artificially intelligent group will require a well-defined metric in order to evaluate the decision problem in such terms. If a decision problem is sufficiently hard, an artificially intelligent decision maker may be motivated to deploy additional sensors and actuators, call for reinforcements, modify its preferences, seek further instructions, and so forth.

The concept of hardness is especially important with mixed-motive multi-stakeholder decision scenarios, where opportunities for both competition and cooperation exist. The Prisoner's Dilemma game is perhaps the best-known mixed-motive example of a two-agent decision problem that involves opportunities for both cooperation and conflict. The essence of the game, and one of the reasons for its enduring interest, is that it serves as a model of situations where mutual cooperation is better than mutual noncooperation for both players, but unilaterally attempting to cooperate leaves one vulnerable to exploitation by the other. This game possesses a great capacity for cooperative behavior, but the dilemma arises because mutual defection is the unique Nash equilibrium, and, according to the doctrine of individual rationality, both players should adopt this pessimistic next-worst solution rather than the more optimistic next-best cooperative solution. Thus, the opportunity costs for mutual defection (the payoffs for mutual cooperation) are actually larger than the equilibrium payoffs.

The classical formulation of the PD game, as illustrated in the payoff matrix displayed in Table 1.1, is completely symmetrical: Both players are given exactly the same payoffs. With ordinal payoffs, where scale does not matter, the symmetrical formulation may be justified. But if the payoffs were expressed in cardinal form, as they would be in realistic PD-type scenarios, there is no reason to expect the two players to function according to the same utility scale. For example, consider the asymmetrical payoff structure illustrated in Table 4.4. The opportunity cost for defecting (the reward for cooperating) is nearly as large as the reward for defecting, regardless of what the other player does. Furthermore, there is a significant opportunity, with little cost, for X_2 to be altruistic by sacrificing some payoff to improve the payoff to X_1. With this payoff scenario, both players might possess some motivation to cooperate. Nevertheless, individual rationality dictates that each player should do what is best for itself and does not account for any propensity for behavior other than seeking the equilibrium solution. Thus, although the structure of the payoffs is not amenable to intrinsic coordination, the possibility still remains that the players may have sufficient knowledge and introspection to achieve extrinsic coordination.

The story lines that accompany PD-type games often offer significant opportunities for the characterization of social concepts such as a willingness to

Table 4.4. *A Prisoner's Dilemma*
scenario with asymmetrical
cardinal utilities

	X_2	
X_1	C	D
C	(90, 44)	(20, 45)
D	(95, 34)	(25, 35)

cooperate and a propensity to exploit. Although these concepts are not represented by the raw payoffs, they often cohere well with observed behavior, as born out by Wolf and Shubik (1974), who describe laboratory experiments involving randomly selected humans that result in the cooperative solution being chosen relatively frequently with single play and no communication between the participants. This evidence suggests that people are often motivated by considerations other than raw payoffs when confronted with such scenarios. There are often expectations of cooperation and nonexploitation and, as the example in Section 3.6 illustrates, the utilities for the PD game can be augmented with parameters that introduce a social component into the game while retaining its essential features. One of the benefits of such an augmented model is that these social parameters may permit the formulation of a quantitative measure of the difficulty, tension, or hardness involved in making decisions.

4.3 Hardness for single agents

The classical categorical utilities paradigm simply assumes that any tension, perplexity, and anxiety associated with the outcomes have been appropriately incorporated into the utilities that represent the preferences of the decision makers. If there is a "hard" part of the problem, therefore, it is in defining the utilities. Once they are defined, the decision problem becomes a dispassionate search to identify the best solution. Thus, the difficulty of making a decision is not a real issue – each decision maker simply does the best thing possible from its point of view. The main remaining challenge associated with the problem is the computational difficulty of conducting the search.

However, the use of conditional utilities permits the participants to define their preferences as functions of the conjectures (hypothetical preferences) of others, thereby incorporating social relationships directly into the decision problem, in contradistinction to Friedman's (1961) proposed division of labor. These social relationships represent the attitudes of decision makers with respect to each others' preferences, and therefore may indeed reflect the tension or anxiety inherent in the relationships. Consequently, the intrinsic difficulty of making a decision may no longer be a moot issue.

As mentioned earlier, one obvious manifestation of hardness is associated with opportunity cost. Opportunity cost, however, is not necessarily a function of the quality of an alternative. To illustrate, suppose X chooses from a set of highly desirable luxury automobiles. If the utilities of the top two candidates are nearly equal, then it is easy to see that the opportunity cost is maximum, even though both choices are good ones. But suppose X is choosing from a set of old, broken-down used cars. Again assuming that the utilities of the top two are nearly equal, the opportunity cost is also maximum, even though all of the choices are bad ones. At the other extreme, suppose X is choosing from a set that contains only one highly desirable car, in which case the opportunity cost is small and the decision is an easy one.

4.3.1 Subordination

Our desire is to develop a mathematical definition of hardness that applies to both single- and multiple-stakeholder decision problems. We approach this issue by first examining the single-stakeholder case, and introduce a measure of how much an alternative is *subordinated* in the sense that it is inferior to the ideal case that would obtain if it were the only alternative with positive utility. For this development, we assume that $\mathcal{A}$ is a finite set of distinct alternatives, utility is nonnegative ($u(a) \geq 0$) and sums to unity ($\sum_a u(a) = 1$). Let $S_u : \mathcal{A} \to \mathbb{R}$ denote the subordination function corresponding to the utility u. We propose that a meaningful concept of subordination should possess the following properties.

1. If $u(a) = 1$, the superiority of a is absolute, since the utility of all other alternatives is zero. Consequently, $S_u(a) = 0$, meaning that a is not subordinated to any other conceivable alternative.
2. If $u(a) < 1$, then $S_u(a) > 0$; that is, the subordination of a should be strictly positive, since that alternative is inferior to a hypothetical alternative with unity utility.
3. Subordination should be strictly monotonically decreasing with respect to $u(a)$; that is, if $u(a) > u(a')$, then $S_u(a) < S_u(a')$.
4. If $u(a) = 0$, then the inferiority of a is absolute; hence, the subordination associated with a should be infinite.

The last condition deserves some comment. Suppose $u(a) = 0$ but $S_u(a) < \infty$. Then we could conceive of an alternative a' such that $S_u(a) < S_u(a')$. But that

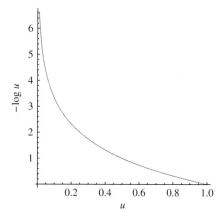

Figure 4.1. The range of the subordination function.

would be impossible, since the utility of a is minimal and subordination is strictly monotonic – a contradiction. Thus, $u(a) = 0$ implies $S_u(a) = \infty$.

A function that complies with these desirable properties is

$$S_u(a) = -\log u(a). \tag{4.1}$$

This function is certainly not the only one that satisfies the above conditions. For example, the function $S'_u(a) = \frac{1}{u(a)} - 1$ also complies. As we proceed with our development, however, we shall see that the function defined by (4.1) possesses intuitively pleasing properties that S'_u lacks.

Figure 4.1 displays a plot of the range of the subordination function over the interval $u \in [0,1]$. Subordination is a measure of the penalty incurred for choosing an alternative a rather than choosing a (possibly hypothetical) alternative upon which all of the utility is concentrated. We see that subordination changes slowly as $u \to 1$ but increases more and more rapidly as $u \to 0$.

To illustrate the use of this function, let $\mathcal{A} = \{a_1, a_2, a_3, a_4\}$, and consider the following three scenarios:

Scenario 1: $u_1(a_1) = 1$, $u_1(a_2) = 0$, $u_1(a_3) = 0$, $u_1(a_4) = 0$.
Scenario 2: $u_2(a_1) = 0.90$, $u_2(a_2) = 0.07$, $u_2(a_3) = 0.02$, $u_2(a_4) = 0.01$.
Scenario 3: $u_3(a_1) = 0.27$, $u_3(a_2) = 0.26$, $u_3(a_3) = 0.24$, $u_3(a_4) = 0.23$.

For Scenario 1,

$$S_{u_1}(a_1) = -\log 1 = 0$$
$$S_{u_1}(a_i) = -\log 0 = \infty, i = 2, 3, 4,$$

which means that a_1 is subordinated to no alternatives, and all other alternatives are subordinated to every alternative (including hypothetical ones) with positive utility. Essentially, the subordination function tells us that choosing a_i, $i \in \{2,3,4\}$ is infinitely inferior to choosing a_1.

For Scenario 2,[1]

$$S_{u_2}(a_1) = -\log 0.9 \approx 0.15$$
$$S_{u_2}(a_2) = -\log 0.07 \approx 3.8$$
$$S_{u_2}(a_3) = -\log 0.02 \approx 5.6$$
$$S_{u_2}(a_4) = -\log 0.01 \approx 6.6.$$

The fact that the subordination of a_1 is not zero, even though it is the highest-ranking alternative, means that it is hypothetically possible to define an alternative with greater utility and, hence, with smaller subordination. Even though the utility of a_i, $i \neq 1$, is much less than the utility of a_1, the decision maker can gain at least some benefit by choosing, say, a_3. Such a choice is clearly not best, and the subordination function provides a measure of the degree of inferiority with respect to an ideal situation (such as Scenario 1).

For Scenario 3,

$$S_{u_3}(a_i) \approx -\log 0.25 = 2, \ i = 1,\dots,4.$$

Since the utility of all alternatives are nearly equal, they all possess nearly the same subordination.

4.3.2 Ambivalence

Utility and subordination enjoy an inverse relationship with each other. When utility is high, subordination is low and vice versa. This property motivates the definition of *ambivalence*, $A_u(a)$, as the product of utility and subordination:

$$A_u(a) = u(a)S_u(a) = -u(a)\log u(a).$$

Figure 4.2 displays a plot of the range of the ambivalence function.

We interpret ambivalence as a measure of the questionableness or vagueness associated with either selecting or not selecting an alternative. Suppose $u(a) \approx 1$. Then $S_u(a) \approx 0$, and the product will be small, a condition of low ambivalence. In other words, since the utility of a is high, the decision maker would be *reluctant not to select a*. Now suppose $u(a) \approx 0$. Then $S_u(a)$ will be large, but the product will be small, again resulting in a condition of low ambivalence. In this

---

[1] In this discussion, all logarithms are with respect to base 2.

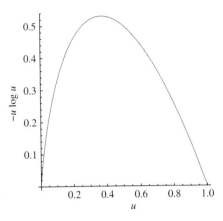

Figure 4.2. The range of the ambivalence function.

case, however, since the utility is low, this means that the decision maker would be *reluctant to select a*. Thus, we see that ambivalence is small for extreme values of the utility, a result that is intuitively reasonable. (These interpretations would not be possible if S'_u were used as the definition of subordination.)

For Scenario 1,

$$A_{u_1}(a_1) = 1\log(1) = 0$$
$$A_{u_1}(a_i) = 0\log 0 = 0, \; i = 2,3,4,$$

where this last equality follows from L'Hopital's rule. Thus, all alternatives have the same ambivalence, namely, zero – there is no vagueness in the choice associated with this decision problem – a_1 is unquestionably good, and a_i, $i \in \{2,3,4\}$ are unquestionably bad.

For Scenario 2,

$$A_{u_2}(a_1) = -0.9\log 0.9 \approx 0.13$$
$$A_{u_2}(a_2) = -0.07\log 0.07 \approx 0.27$$
$$A_{u_2}(a_3) = -0.02\log 0.02 \approx 0.11$$
$$A_{u_2}(a_4) = -0.01\log 0.01 \approx 0.07.$$

The properties of the ambivalence function are clearly seen by comparing $A_{u_2}(a_1)$ and $A_{u_2}(a_3)$. Action a_1 has a very high utility value and a low subordination value, while a_3 has a low utility value and a high subordination value. These values indicate, on the one hand, that there is little vagueness associated with selecting a_1, and, on the other hand, there is little vagueness associated with not selecting a_3 (and even less vagueness associated with not selecting a_4).

For Scenario 3,

$$A_{u_3}(a_i) \approx -0.25 \log 0.25 = 0.5, \ i = 1,\ldots,4,$$

which corresponds to the obvious fact, with this scenario, that there is little difference in the vagueness associated with selecting any of the alternatives.

4.3.3 Hardness

Selecting or rejecting an action with high ambivalence is not as satisfying as selecting or rejecting one with low ambivalence. Intuitively, high-ambivalence actions are more difficult than low-ambivalence ones. Consequently, a meaningful definition of the overall difficulty associated with making a choice is the total ambivalence associated with the actions. Accordingly, let us define *hardness* as the sum of the ambivalence of each of the actions, yielding

$$H[u(a_1),\ldots,u(a_n)] = \sum_i A_u(a_i) = -\sum_i u(a_i) \log u(a_i).$$

Obviously, if all of the utility is concentrated on one action, then hardness is minimal. What constitutes a maximally hard decision? To answer this question, we use a Lagrange multiplier to account for the constraint that the sum of all utility values must be unity. Accordingly, define

$$J(u_1,\ldots,u_n) = -\sum_i u_i \log u_i + \lambda \sum_i u_i.$$

Taking the derivative of J with respect to u_i and setting the result to zero yields

$$\frac{\partial J(u_1,\ldots,u_n)}{\partial u_i} = -\log u_i - 1 + \lambda = 0$$

for $i = 1,\ldots,n$. Thus, all of the u_i's must be equal, and since their sum must be unity, we conclude that ambivalence of a_i is maximized when $u(a_i) = \frac{1}{n}$, $i = 1,\ldots,n$. Thus, a maximum hardness scenario obtains when all alternatives have exactly the same utility, in which case,

$$H\left(\tfrac{1}{n},\ldots,\tfrac{1}{n}\right) = \sum_i \frac{1}{n} \log n = \log n.$$

Figure 4.3 displays $H(u, 1-u)$ for the case $n = 2$. We see that $H(u, 1-u) \approx 0$ for $u \approx 1$ and $u \approx 0$, and $H(u, 1-u)$ achieves is maximum value for $u = \frac{1}{2}$.

For Scenario 1, $H[u(a_1),\ldots,u(a_4)] = 0$, indicating a complete lack of hardness. Making a decision under this scenario is trivial. For Scenario 2,

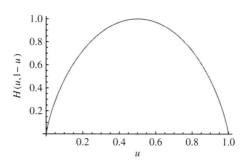

Figure 4.3. The hardness function for $n = 2$.

$H[u(a_1),\ldots,u(a_4)] \approx 0.58$, indicating a modest amount of hardness, again resulting in an easy decision with low opportunity cost. For Scenario 3, $H[u(a_1),\ldots,u(a_4)] = 1.997$, indicating near maximum hardness – a decision with a near maximum opportunity cost.

The function H is traditionally called *entropy*. The word comes from the Greek and means "transformation." In the nineteenth century, the German physicist Rudolf Clausius introduced the term to describe the tendency for the world to move to a state of increasing disorder (the second law of thermodynamics). The concept was refined by Ludwig Boltzmann, who introduced the logarithmic mathematical model of the concept. In the thermodynamics context, entropy is a measure of the inability to convert energy into work (Prigogine, 1955; Georgescu-Roegen, 1971; Bailey, 1990).

In the mid-twentieth century, the engineer/mathematician Claude Shannon developed a mathematical model of information theory and employed the concept of entropy as a measure of the degree of randomness associated with a phenomenon, such as a message transmitted through an imperfect medium (Shannon and Weaver, 1949). Put in vernacular terms, entropy is a measure of the intrinsic capacity to make a wrong guess. To illustrate, let us recast these three scenarios in an epistemological context and suppose that instead of viewing the a_i's as alternatives for a decision maker, we view them as propositions such that one, and only one, of them is true. In that context, $u(a_i)$ may be interpreted as the probability that a_i is true. For Scenario 1, since the probability that a_1 is true is one, there is zero capacity to guess wrong (assuming, of course, that the decision maker is rational and acts according to the information in its possession). For Scenario 2, guessing that a_1 is true will most likely be correct; hence, the capacity to guess wrong is small. For Scenario 3, where all of the propositions are nearly equally likely, the capacity to guess wrong is near maximum.

In the information-theoretic context, the function S_u is sometimes called the *surprise* function (see Ross, 2002). This terminology arises from the intuitively meaningful notion that an individual would be greatly surprised if an extremely unlikely event were to be true. Thus, the praxeological notion of subordination is analogous to the epistemological notion of surprise.

Relating the epistemological context of entropy to the praxeological context of hardness, the next most probable outcome is analogous to opportunity cost. Shannon's information theory, originally conceived as a way to characterize the intrinsic properties of a communications system in the presence of uncertainty, also has application as a way to characterize the intrinsic properties of a decision problem, where we are concerned with the praxeological concept of the difficulty in making decisions rather than the epistemological concept of uncertainty. Thus, we employ the mathematical syntax of Shannon's information theory, but with praxeological, rather than epistemological, semantics. Equating the concept of hardness in the praxeological domain to the notion of entropy in the epistemological domain is another illustration of the appropriation of concepts that were originally developed in one application domain for another domain.[2]

4.4 Mutual information and coordination

In the traditional communications context, entropy is used to form a notion of what is called *channel capacity*, or the rate at which information can be transmitted through an imperfect communications medium with controllably small probability of error. In this regard, the key attribute involved in defining this capacity is the so-called mutual information of the transmitter and receiver. Essentially, the channel capacity is obtained by maximizing the mutual information over all input distributions. This notion of capacity, however, does not have an obvious application to the praxeological issue that we are addressing. Nevertheless, the notion of mutual information, as we shall develop it, does have relevance to the problem of ascertaining the ability of a group to coordinate.

4.4.1 Joint Hardness

The key concept in the development of mutual information is the idea of joint hardness. Hardness, however, is a more general concept than opportunity cost. When a group of decision makers must function in a social environment, the

---

[2] It should be noted that Shannon's information theory has been employed in a wide variety of applications, including feature extraction in machine learning applications, image registration, hierarchical clustering, and so forth.

concept of what is best, what is second best, and so forth, depends upon the perspectives of the individual participants. In a society where the interests of the stakeholders are diametrically opposed, each participant will seek to maximize its personal benefit. In such a case, although the individuals are competing, they are behaving in a highly coherent way in that their behavior reveals a sense of understandability and logical interconnectedness. At another extreme, consider a society in which the interests of all individuals perfectly coincide. Here, the propensity to behave coherently is also high. Although different in structure, these two decision scenarios are, generally speaking, easy ones. Between the two extremes of pure competition and pure cooperation, however, lies a large class of mixed-motive scenarios, where opportunities for competition, cooperation, and compromise exist to varying degrees. With such scenarios, the ability of the society to behave coherently is not so obvious. Thus, in a multi-stakeholder context, hardness must be associated with the ability of the group to function coherently as well as with individual opportunity costs.

Notational convention Although subordination, ambivalence, and hardness are all defined in terms of utilities, it is often convenient to express these terms as functions of the stakeholders. Thus, for X with utility u_X, the notations $H(X)$ and $H(u_X)$ will be taken to mean the same thing and will be used interchangeably.

Definition 4.1 Let $\mathcal{X}_n = \{X_1, \ldots, X_n\}$ be a group of n stakeholders and let $v_{X_1 \cdots X_n}$ be a group welfare function as developed in Chapter 3 (see (3.4)). Also, let v_{X_i} denote the individual welfare function defined in Chapter 3 (see (3.5)). The *joint hardness* of the group $\mathcal{X}_n$ is

$$H(X_1, \ldots, X_n) = - \sum_{a_1, \ldots, a_n} v_{X_1 \cdots X_n}(a_1, \ldots, a_n) \log v_{X_1 \cdots X_n}(a_1, \ldots, a_n).$$

□

In particular, for $n = 2$, let $\mathcal{A}_1 = \{a_{11}, \ldots, a_{1k}\}$ and $\mathcal{A}_2 = \{a_{21}, \ldots, a_{2m}\}$ denote the action spaces for X_1 and X_2, respectively. Then

$$H(X_1, X_2) = - \sum_{i,j} v_{X_1 X_2}(a_{1i}, a_{2j}) \log v_{X_1 X_2}(a_{1i}, a_{2j})$$

and

$$H(X_1) = - \sum_i v_{X_1}(a_{1i}) \log v_{X_1}(a_{1i})$$

$$H(X_2) = - \sum_j v_{X_2}(a_{2j}) \log v_{X_2}(a_{2j}).$$

We may also define joint hardness in the social choice context (see Section 3.9) as

$$H(X_1, \ldots, X_n) = -\sum_{a_1, \ldots, a_n} W_{X_1 \cdots X_n}(a_1, \ldots, a_n) \log W_{X_1 \cdots X_n}(a_1, \ldots, a_n),$$

where $W_{X_1 \cdots X_n}$ is the social choice concordant utility defined by (3.7), and the *individual hardness* for each X_i is

$$H(X_i) = -\sum_{a_i} w_{X_i}(a_i) \log w_{X_i}(a_i),$$

where w_{X_i} is the individual welfare function defined by (3.9).

We state the following theorem (for a proof, see Cover and Thomas, 1991):

Theorem 4.1

$$H(X_1, \ldots, X_n) \geq \max\{H(X_1), \ldots, H(X_n)\}$$

and

$$H(X_1, \ldots, X_n) \leq H(X_1) + \cdots + H(X_n),$$

with equality if $X_1, \ldots, X_n$ are mutually praxeologically independent.

4.4.2 Conditional hardness

Thus far in our development, we have provided definitions for the hardness of an individual stakeholder and for a group of stakeholders viewed collectively. We now consider the notion of hardness for a subgroup conditioned on the conjectures of another subgroup. In the interest of clarity and brevity, we introduce this topic by initially restricting $n = 2$. To address this issue, we first compute the hardness of the conditional utility of X_2 given that X_1 conjectures a_1, yielding

$$H(X_2 | X_1 \to a_1) = -\sum_{a_2} v_{X_2|X_1}(a_2|a_1) \log v_{X_2|X_1}(a_2|a_1),$$

where

$$v_{X_2|X_1}(a_2|a_1) = \frac{v_{X_1 X_2}(a_1, a_2)}{v_{X_1}(a_1)} \tag{4.2}$$

and the notation $X_1 \to a_1$ denotes the conjecture by X_1 of a_1. (Note: In the social choice context, the group welfare function $v_{X_1 X_2}$ in (4.2) is replaced by the social choice concordant utility $W_{X_1 X_2}$ defined by (3.7) and the individual welfare function v_{X_1} is replaced by w_{X_1}.)

The function $H(X_2|X_1 \to a_1)$ represents X_2's cumulative ambivalence given X_1's conjecture. To compute the conditional hardness of X_2 given X_1, we sum the product of this cumulative ambivalence, weighted by the individual welfare of X_1, over all $a_1 \in \mathcal{A}_1$, yielding

$$H(X_2|X_1) = \sum_{a_1} v_{X_1}(a_1) H(X_2|X_1 \to a_1)$$

$$= -\sum_{a_1} v_{X_1}(a_1) \sum_{a_2} v_{X_2|X_1}(a_2|a_1) \log v_{X_2|X_1}(a_2|a_1)$$

$$= -\sum_{a_1} \sum_{a_2} v_{X_1 X_2}(a_1,a_2) \log v_{X_2|X_1}(a_2|a_1).$$

We may interpret conditional hardness as a measure of how much a conjecture by X_1 does *not* reduce the hardness for X_2. Intuitively, it is clear that $H(X_2|X_1) \le H(X_2)$. It is straightforward to see that if X_1 and X_2 are praxeologically independent, then $H(X_2|X_1) = H(X_2)$; that is, a conjecture by X_1 does not change X_2's hardness. At the other extreme, if, for each a_1 there exists a unique a_2 such that $v_{X_2|X_1}(a_2|a_1) = 1$, then $H(X_2|X_1) = 0$, indicating the maximum reduction in hardness for X_2 given a conjecture by X_1. In particular, it is obvious that $H(X_1|X_1) = 0$.

We state without proof the following results, which establish that the hardness associated with a joint decision problem is the hardness associated with one plus the conditional hardness of the other.

Theorem 4.2 *The chain rule of hardness:*

$$H(X_1, X_2) = H(X_1) + H(X_2|X_1).$$

Theorem 4.3

$$H(X_1, X_2|X_3) = H(X_1|X_3) + H(X_2|X_1, X_3).$$

For proofs of these results, see Cover and Thomas (1991).

4.4.3 Relative hardness and mutual information

Suppose u and v are two possible utilities for a decision problem. The *relative hardness* is a measure of the distance, or dissimilarly, between the two utilities, and is defined as

$$D(u\|v) = \sum_i u(a_i) \log \frac{u(a_i)}{v(a_i)}.$$

Important properties of relative hardness are summarized by the following theorem (For a proof, see Cover and Thomas, 1991).

Theorem 4.4 *Let u and v be two utilities defined over $\mathcal{A}$. Then*

$$D(u\|v) \geq 0$$

with equality if and only if $u(a) = v(a)$ for all $a \in \mathcal{A}$.

Relative hardness, traditionally termed the *Kullback–Liebler distance*, is not a true metric. It is not symmetric, and it does not obey the triangle inequality. Nevertheless, it is often convenient to think of relative hardness as a "distance" between utilities (we will subsequently refine this concept).

Definition 4.2 Let $\mathcal{X}_n = \{X_1, \ldots, X_n\}$ be a group of n stakeholders with group welfare function $v_{X_1 \cdots X_n}$ and individual welfare functions $v_{X_1}, \ldots, v_{X_n}$. The *mutual information* is the Kullback–Liebler distance between the group welfare function and the product of the individual welfare functions.

$$
\begin{aligned}
I(X_1, \ldots, X_n) &= D(v_{X_1 \cdots X_n} \| v_{X_1}, \ldots, v_{X_n}) \\
&= \sum_{a_1} \cdots \sum_{a_n} v_{X_1 \cdots X_n}(a_1, \ldots, a_n) \log \frac{v_{X_1 \cdots X_n}(a_1, \ldots, a_n)}{v_{X_1}(a_1) \cdots v_{X_n}(a_n)} \quad (4.3) \\
&= \sum_i H(X_i) - H(X_1, \ldots, X_n).
\end{aligned}
$$

$\square$

From Theorem 4.1 we see that $I(X_1, \ldots, X_n) = 0$ if $X_1, \ldots, X_n$ are mutually praxeologically independent. This is the situation that obtains when all utilities are categorical.

Conventional applications of Shannon information theory typically involve only two stakeholders, in which case mutual information is

$$I(X_1, X_2) = H(X_1) - H(X_2|X_1) = H(X_1) + H(X_2) - H(X_1, X_2).$$

Also of interest is the *self-information* associated with a single stakeholder:

$$I(X_1, X_1) = H(X_1).$$

Mutual information in the traditional probabilistic context is a measure of the amount of information the random variables share. For the two-random variable case, if one of them is known, the mutual information is a measure of how much uncertainty is reduced about the other random variable. If the two random variables are independent, then mutual information is zero, and mutual information is maximized if there is a one-to-one relationship between the values that the two random variables assume.

In the praxeological domain, where utilities, rather than probabilities, are involved, mutual information becomes a measure of the amount of interest that the stakeholders share. For the two-stakeholder case, if one takes an action, the mutual information is a measure of how much hardness is reduced for the other stakeholder. if the two stakeholders are praxeologically independent (i.e., all utilities are categorical), then mutual information is zero, and mutual information is maximized if there is a one-to-one relationship between their actions (i.e., they are slaved together).

To make this analysis explicit, consider the argument of the logarithm term in (4.3). The numerator $v_{X_1 \cdots X_n}(a_1, \ldots, a_n)$ measures the welfare of the group considered as an entity, if the action profile $(a_1, \ldots, a_n)$ is implemented, and the denominator $v_{X_1}(a_1) \cdots v_{X_n}(a_n)$ is the product of the measures of individual welfare. If the stakeholders are praxeologically independent, then the numerator is equal to the denominator, the ratio is unity, and, consequently, the logarithm is zero.

4.4.4 Coordination capacity for two stakeholders

Mutual information provides a powerful framework within which to examine the ability of two stakeholders to coordinate their behavior. In the extreme case of praxeological independence, where the interests of one stakeholder are not influenced by the interests of the other, then, in terms of the utility structure only (ignoring psychological and sociological considerations that are not encoded into the payoffs), there is no capacity for the players to coordinate, although they may cooperate if their interests just happen to coincide. Each is free to do the best for itself without allowing concern for the welfare of the other to influence its decision. Accordingly, the mutual information is zero.

On the other hand, if the stakeholders are slaved together, such that the behavior of one completely constrains the behavior of the other, then the two are maximally coordinated. Between these two extremes, mutual information may serve to define the intrinsic capacity of the two stakeholders to coordinate. To proceed with this development, it is convenient first to refine the Kullback–Liebler distance. Although it is a measure of the relative hardness of the two stakeholders, the Kullback–Liebler distance measure is not a true metric, as discussed earlier. The following theorem, however, does define a metric between the utilities of two stakeholders.

Theorem 4.5 *The function*

$$d(X_1, X_2) = H(X_1, X_2) - I(X_1, X_2)$$

possesses the following properties:

$d(X_1, X_2) \geq 0$ *and* $d(X_1, X_2) = 0$ *if and only if* $X_1 = X_2$

$d(X_1, X_2) = d(X_2, X_1)$ *(symmetry)*

$d(X_1, X_2) \leq d(X_1, X_3) + d(X_3, X_2)$ *(triangle inequality),*

where the notation $X_1 = X_2$ *means that there is a one-to-one function mapping the actions of* X_1 *to the actions of* X_2*; that is,* X_1 *takes action* $a_1 \in \mathcal{A}_1$ *if and only if* X_2 *takes action* $a_2 \in \mathcal{A}_2$.

For a proof of this theorem, see Li et al. (2001) and Kraskov et al. (2003). Two equivalent expressions for $d(\cdot, \cdot)$ are

$$d(X_1, X_2) = 2H(X_1, X_2) - H(X_1) - H(X_2)$$

and

$$d(X_1, X_2) = H(X_1|X_2) + H(X_2|X_1).$$

We shall refer to $d(X_1, X_2)$ as a *dispersion measure*. This measure is maximum when X_1 and X_2 are praxeologically independent, in which case it equals the total hardness of the group: $d(X_1, X_2) = H(X_1) + H(X_2)$ (Kraskov et al., 2003). As pointed out in Li et al. (2001) and Kraskov et al. (2003), a more useful metric is obtained by dividing $d(X_1, X_2)$ by the joint hardness, yielding

$$\mathcal{D}(X_1, X_2) = 1 - \frac{I(X_1, X_2)}{H(X_1, X_2)} = \frac{d(X_1, X_2)}{H(X_1, X_2)}.$$

This function measures the *relative dispersion*[3] between the two utilities. It is clearly symmetric and nonnegative by inspection, and the proof that $\mathcal{D}$ satisfies the triangle inequality is also provided in Li et al. (2001) and Kraskov et al. (2003).

Finally, we define the *coordination capacity* as

$$C(X_1, X_2) = 1 - \mathcal{D}(X_1, X_2).$$

The coordination capacity is a measure of the degree to which X_1 and X_2 are socially connected and serves as a measure of the intrinsic ability of the stakeholders to coordinate their behavior independently of the rationale used to define a solution. When all of the utilities are categorical, as is the case with classical games, the stakeholders are mutually praxeologically independent and the relative dispersion is maximum, that is, $\mathcal{D}(X_1, X_2) = 1$. Consequently, the coordination capacity is zero. This result does not mean, of course, that the

---

[3] This measure is similar in concept to the *Jaccard index*, or *coefficient de communauté*, a statistic for comparing the similarity and diversity of sample sets (Jaccard, 1901).

mutually praxeologically independent players cannot coordinate extrinsically according to notions of rational behavior as discussed in Section 4.1. Nor does a lack of *capacity* to coordinate intrinsically mean that the agents will not function harmoniously. Rather, it means that if they do, it is simply by coincidence, based on the juxtaposition of their utilities, and not because of a social relationship between them.

As discussed in Section 4.1, extrinsic coordination focuses on the propensity of a group to reinforce cooperative behavior, and thus has a constructive connotation. However, intrinsic coordination, which focuses on the structure of the utilities, has a neutral connotation and deals with both constructive and destructive behavior. The following example illustrates both types of coordination.

Example 4.1 Let us return to Example 2.2, which considers two agents, X_1 and X_2 facing each other on opposite sides of a doorway just wide enough for two people to pass through. The actions available to both agents are to vere to the right (r) or to the left (l). We assume that both X_1 and X_2 are completely dissociated. Let us suppose that X_1 is not able to anticipate any moves by X_2, and thus X_1 possess a uniform categorical utility of the form

$$u_{X_1}(r) = u_{X_1}(l) = \frac{1}{2}.$$

X_2, however, is able to condition its utility to account for anticipated behavior of X_1. We consider two scenarios.

Cooperative scenario X_2 possesses a conditional utility that is designed to allow both agents to pass through the doorway; hence,

$$u_{X_2|X_1}(r|r) = 1 \quad u_{X_2|X_1}(l|r) = 0$$
$$u_{X_2|X_1}(r|l) = 0 \quad u_{X_2|X_1}(l|l) = 1.$$

Thus, if X_1 conjectures r, then X_2 will also place all of its conditional utility on r, thus allowing both to pass, similarly if X_1 conjectures l, then X_2 will also place its conditional utility mass on l. Aggregating these utilities yields the group welfare function

$$v_{X_1 X_2}(r,r) = \frac{1}{2} \quad v_{X_1 X_2}(r,l) = 0$$
$$v_{X_1 X_2}(l,r) = 0 \quad v_{X_1 X_2}(r,l) = \frac{1}{2}.$$

By straightforward calculations, $H(X_1) = H(X_2) = H(X_1, X_2) = 1$, $I(X_1, X_2) = 1$, $d(X_1, X_2) = 0$ and, consequently, $C(X_1, X_2) = 1$, a condition of maximum coordination capacity in the constructive sense.

Conflictive scenario X_2 possesses a conditional utility that is designed to block X_1's progress, hence,

$$u_{X_2|X_1}(r|r) = 0 \quad u_{X_2|X_1}(l|r) = 1$$
$$u_{X_2|X_1}(r|l) = 1 \quad u_{X_2|X_1}(l|l) = 0.$$

Thus, if X_1 conjectures r, then X_2 will place all of its conditional utility on l, thus blocking X_1. Aggregating these utilities yields the group welfare function

$$v_{X_1 X_2}(r,r) = 0 \quad v_{X_1 X_2}(r,l) = \frac{1}{2}$$
$$v_{X_1 X_2}(l,r) = \frac{1}{2} \quad v_{X_1 X_2}(r,l) = 0.$$

By straightforward calculations, $H(X_1) = H(X_2) = H(X_1, X_2) = 1$, $I(X_1, X_2) = 1$, $d(X_1, X_2) = 0$ and, consequently, $C(X_1, X_2) = 1$, also a condition of maximum coordination capacity, but in the destructive sense. □

Example 4.2 We now compute the coordination capacity for the Prisoner's Dilemma game defined in Section 3.6. The corresponding joint and individual hardness values are

$$H(X_1, X_2) = -\alpha \log \alpha - \alpha\beta \log \alpha\beta - (1 - \alpha - \alpha\beta) \log(1 - \alpha - \alpha\beta), \quad (4.4)$$

$$H(X_1) = -\alpha \log \alpha - (1 - \alpha) \log(1 - \alpha), \quad (4.5)$$

$$H(X_2) = -(\alpha + \alpha\beta) \log(\alpha + \alpha\beta) - (1 - \alpha - \alpha\beta) \log(1 - \alpha - \alpha\beta). \quad (4.6)$$

Figure 4.4 displays contour plots of $H(X_1, X_2)$ and $I(X_1, X_2)$. We see that hardness is maximized in the region where $\alpha \approx \beta \approx \frac{1}{2}$ and is zero when $\alpha = 0$. Also, mutual information is maximized when $\beta = 0$ and $\alpha \approx \frac{1}{2}$. Plots of $d(X_1, X_2)$ and $C(X_1, X_2)$ are displayed in Figure 4.5. The coordination capacity is maximum when the exploitation index β is zero and is minimum when $\alpha + \beta \approx 1$. Of particular interest is the value of the index in the (β, α) triangular-shaped region corresponding to the PD game, as indicated on the plot. The coordination capacity in this region lies in the interval $(0.2, 0.35)$, which indicates a nonnegligible capacity to coordinate. This example demonstrates that the use of conditional utilities allows us to remain faithful to the essence of the PD game in the sense that the ordinal relationships among the profiles is consistent with

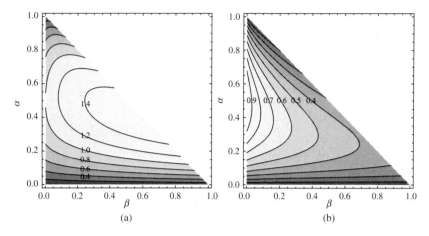

Figure 4.4. Contour plot of (a) $H(X_1, X_2)$ and (b) $I(X_1, X_2)$ for the Prisoner's Dilemma game.

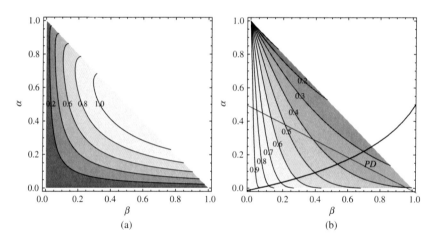

Figure 4.5. Contour plot of (a) $d(X_1, X_2)$ and (b) $C(X_1, X_2)$ for the Prisoner's Dilemma game.

the traditional ordinal PD utility structure. With this formulation, however, we see that the capacity to coordinate is nonzero, which is in contrast to a coordination capacity of zero with the traditional formulation of the game in terms of categorical utilities. The major difference between the two formulations is that the conditional formulation imposes a hierarchical structure on the stakeholders; that is, X_2's preferences are conditioned on X_1's preferences. This example provides an instance where the social attitudes of the players can influence, and even override, the purely selfish tendency to take advantage of one who leaves itself vulnerable to exploitation in an attempt to cooperate. □

4.4.5 Coordination capacity for $n > 2$ stakeholders

When $n \geq 3$, we may expand the definition of $d(X_1, \ldots, X_n)$ to become

$$d(X_1, \ldots, X_n) = (n-1)H(X_1, \ldots, X_n) - I(X_1, \ldots, X_n)$$

$$= nH(X_1, \ldots, X_n) - \sum_{i=1}^{n} H(X_i).$$

This function is symmetric and nonnegative and is zero if and only if there is a one-to-one function mapping X_i to X_j for all i, j. As with the case $n = 2$, $d(X_1, \ldots, X_n)$ is a *dispersion measure*; that is, a measure of how much the utilities of the group are in conflict. $d(X_1, \ldots, X_n)$ achieves its maximum when all of the X_i's are mutually praxeologically independent, in which case $d(X_1, \ldots, X_n) = (n-1)\sum_{i=1}^{n} H(X_i)$. We define the *relative dispersion measure* as

$$\mathcal{D}(X_1, \ldots, X_n) = \frac{1}{n-1}\left[1 - \frac{I(X_1, \ldots, X_n)}{H(X_1, \ldots, X_n)}\right]$$

$$= \frac{1}{n-1}\frac{d(X_1, \ldots, X_n)}{H(X_1, \ldots, X_n)}.$$

When $n \geq 3$, $\mathcal{D}(X_1, \ldots, X_n)$, although also not a metric since it involves more than two agents, is symmetric and nonnegative, and equals zero if and only if there is a one-to-one function mapping X_i to X_j for all i, j. It assumes a maximum value of unity when the agents are mutually praxeologically independent.

We define the *coordination capacity* as the difference between unity and the relative dispersion measure, namely,

$$C(X_1, \ldots, X_n) = 1 - \mathcal{D}(X_1, \ldots, X_n).$$

If the stakeholders are mutually praxeologically independent, $\mathcal{D}(X_1, \ldots, X_n) = 1$, the ability to coordinate is minimized, and the optimal behavior is for each stakeholder to do what is best for itself, regardless of the effect on others. This latter case is the operative condition for classical game theory, where all utilities are categorical.

For any k dimensional subgroup $\{X_{i_1}, \ldots, X_{i_k}\}$, $2 \leq k < n$, we may define the corresponding subgroup dispersion measure and relative dispersion measure as

$$d(X_{i_1}, \ldots, X_{i_k}) = (k-1)H(X_{i_1}, \ldots, X_{i_k}) - I(X_{i_1}, \ldots, X_{i_k})$$

$$= kH(X_{i_1}, \ldots, X_{i_k}) - \sum_{j=1}^{k} H(X_{i_j}),$$

and

$$\mathcal{D}(X_{i_1}, \ldots X_{i_k}) = \frac{1}{k-1} \frac{d(X_{i_1}, \ldots, X_{i_k})}{H(X_{i_1}, \ldots, X_{i_k})}.$$

The subgroup coordination capacity is then

$$C(X_{i_1}, \ldots X_{i_k}) = 1 - \mathcal{D}(X_{i_1}, \ldots X_{i_k}).$$

Example 4.3 Larry, Curly, and Moe are going to have a potluck dinner. Larry (L) will bring either salad (sd) or soup (sp); Curly (C), will bring the main course, composed of either beef (bf), chicken (cn), or pork (pk); and Moe (M) will bring dessert, either banana cream pie (bc) or lemon custard pie (lc). Thus, $\mathcal{A}_{X_L} = \{sd, sp\}$, $\mathcal{A}_{X_C} = \{pf, cn, pk\}$, $\mathcal{A}_{X_M} = \{bc, lc\}$.

We assume that Larry's preferences are not influenced by the choice of main course or dessert; thus, he possesses a categorical utility, favoring salad over soup by a factor of 4. His utility is thus $u_{X_L}(sd) = 0.8$ and $u_{X_L}(sp) = 0.2$.

Curly, however, does have discriminating tastes, and his preferences for the main course depend on whether soup or salad is to be served. If salad were served, he would strongly prefer beef to either chicken or pork. However, if soup were served, he would strongly prefer pork over both beef and chicken. Thus, Curly's conditional utilities are

$$u_{X_C|X_L}(bf|sd) = 0.8 \qquad u_{X_C|X_L}(bf|sp) = 0.1$$

$$u_{X_C|X_L}(cn|sd) = 0.1 \qquad u_{X_C|X_L}(cn|sp) = 0.1$$

$$u_{X_C|X_L}(pk|sd) = 0.1 \qquad u_{X_C|X_L}(pk|sp) = 0.8.$$

Finally, Moe also has discriminating tastes, and his preferences for dessert depend on the conjectures of Larry and Curly. If salad and beef, or salad and pork, or soup and pork were served, Moe would strongly prefer banana cream pie; otherwise, he would strongly prefer lemon custard pie. Thus, Moe's conditional utilities are

$$u_{X_M|X_LX_C}(bc|sd,bf) = 1$$
$$u_{X_M|X_LX_C}(lc|sd,bf) = 0$$
$$u_{X_M|X_LX_C}(bc|sd,cn) = 0$$
$$u_{X_M|X_LX_C}(lc|sd,cn) = 1$$
$$u_{X_M|X_LX_C}(bc|sd,pk) = 1$$
$$u_{X_M|X_LX_C}(lc|sd,pk) = 0$$
$$u_{X_M|X_LX_C}(bc|sp,bf) = 0$$
$$u_{X_M|X_LX_C}(lc|sp,bf) = 1$$
$$u_{X_M|X_LX_C}(bc|sp,cn) = 1$$
$$u_{X_M|X_LX_C}(lc|sp,cn) = 1$$
$$u_{X_M|X_LX_C}(bc|sp,pk) = 1$$
$$u_{X_M|X_LX_C}(lc|sp,pk) = 0.$$

Figure 4.6 illustrates the social influence flows among these three agents.

Straightforward calculations reveal that the optimal choice for the group is (salad, beef, banana cream pie), which agrees with the individually optimal choices. Thus, a consensus is achieved. An examination of the hardness and coordination capacity is instructive:

$$H(L,C,M) = 1.655$$
$$I(L,C,M) = 0.820$$
$$d(L,C,M) = 2.490$$
$$C(L,C,M) = 0.248.$$

To calibrate these results, it will be instructive to compare them to the extreme values they could possibly obtain, given the actions available. The maximum

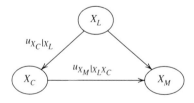

Figure 4.6. The directed acyclic graph for the potluck dinner.

hardness would obtain if the participants were independent and each possessed uniformly distributed categorical utilities, in which case the maximum would be $\log n$, where n is the number of options available to the participant. Since $n = 2$ for Larry and Moe and $n = 3$ for Curly, the maximum hardness is $\log 2 + \log 3 + \log 2 = 3.584$. We thus conclude that this decision problem is of moderate hardness, and also of moderate capacity to coordinate.

It is interesting also to examine the relationships between subgroups as follows:

$$H(L,C) = 1.644$$
$$H(L,M) = 1.244$$
$$H(C,M) = 1.356$$
$$C(L,C) = 0.183$$
$$C(L,M) = 0.008$$
$$C(C,M) = 0.291.$$

These results indicate that all three subgroups have moderately hard decisions to make. They also reveal that the ability for Curly and Moe to coordinate is much greater than the ability of Larry and Moe to coordinate, and somewhat greater than the ability of Larry and Curly to coordinate.

Finally, the individual hardness values for the three agents are

$$H(L) = 0.722$$
$$H(C) = 1.222$$
$$H(M) = 0.532,$$

indicating that Moe has a somewhat easier problem than either Larry or Curly. $\quad\square$

4.5 Summary

Analyzing utility structure with respect to hardness and coordination capacity does not directly help stakeholders make choices. Rather, it gives them an understanding of the ambivalence associated with the choices. It does not tell them what the opportunity costs are, but it does tell them whether those costs are high or low. It provides a quantifiable measure of the intrinsic difficulty in making joint decisions, as well as a measure of the intrinsic ability for the players to coordinate their choices. Consequently, an analysis of hardness and coordination capacity may be viewed as *metaknowledge* that will help the stakeholders more completely to understand just how fit, or adapted, they are to address the decision problem they are facing. In this sense, we may view coordination capacity as an ecological issue that addresses how well a group is able to function in its environment, including relationships with each other. As we have defined it, coordination capacity is zero if all agents possess categorical utilities, regardless of the rationale they use when making choices. The advantage of this notion of coordination capacity is that it removes dependency on motives and other subjective issues from consideration, and focuses on the *structure* of the organization. The disadvantage of this concept is that does not provide the stakeholders with information that is directly useful for making choices.

How might such information be used? Simply recognizing that a stakeholder is facing a challenging task that may consume a great many resources, or will result in high opportunity costs, or is simply making the best of a bad situation, is knowledge that the decision maker can use in several ways. Perhaps the "wisest" action a stakeholder can take is to recognize when it is incapable of making a good choice given its current state of information. When stakeholders recognize that they are not powerful enough, decisive enough, or interconnected enough to make good choices, they may consider the following alternatives.

1. Stakeholders may be motivated to seek help, either in the form of advice, assistance, or therapy. Arrow (1974) observed that "An unsatisfactory solution may be what is needed to provoke the needed information gathering to produce a better one" (p. 48).
2. Stakeholders may be motivated to commit additional resources. For example, an artificially intelligent decision maker (such as a robot) may be prompted to deploy additional sensors or actuators in an attempt to match the environment more effectively.
3. Stakeholders may be motivated to adapt their preferences in an attempt to match the preferences of others. For example, an agent may come to a decision problem with preferences that are not compatible with the preferences of others. If that agent is not intransigent, it may be willing to modify its preferences to effect a compromise. Alternatively, an agent may use measures of hardness and coordination capacity to determine whether or not it enjoys a position of strategic strength or weakness when negotiating a compromise.

5

Uncertainty

Probability theory was to native good sense what a telescope or
spectacles were to the naked eye: a mechanical extension designed
along the same basic principles as the original.
> — *Gerd Gigerenzer et al.*
> *The Empire of Chance (Cambridge University*
> *Press, 1989)*

The question of how to make choices in the presence of uncertainty that arises
because of the lack of complete information has captured the interest of decision
theorists for centuries. The approach to dealing with such situations depends
on how the decision maker characterizes the uncertainty. With a situation of
complete ignorance, it is difficult to proceed. Consequently, much effort has
been devoted to devising rules to govern the way inferences are made. As put by
Gigerenzer et al. (1989), "Although these rules cannot eliminate the uncertainty
intrinsic to the situation, they do eliminate the uncertainty of opinion about how
to weigh what information one has, and therefore, about what course of action
to pursue" (p. 286). The approach that has most widely influenced science
and economics is to comply with the *certainty-equivalence hypothesis*: Given a
decision problem whose outcome is uncertain, it is assumed that a payoff value
exists such that the decision maker is indifferent between receiving that payoff
for certain and the uncertain outcome. A decision maker who complies with
the certainty-equivalence hypothesis will then act as if the certainty-equivalent
payoff were deterministically defined.

Two issues govern how one arrives at such a certainty-equivalent value:
(a) the benefit that accrues to the stakeholder for each possible outcome and
(b) the plausibility of that outcome obtaining. Benefit is typically expressed
with a cardinal utility function that not only orders the outcomes in terms of
preference but also provides a relative measure of the intensity of preference.

126

The most widely used mathematical approach to modeling plausibility is probability theory. When that model is employed, the certainty-equivalent value is defined as the expected value of the utility. The notion of *expected utility* has emerged as the predominant mechanism for making choices when the outcome is uncertain.

The application of probability theory as a mechanism with which to define a rational choice leaves open the choice of which probability model to employ. The strongest possible assertion is what Levi (1980) terms *credal uniqueness*, meaning that there is one, and only one, admissible probability model to describe the random behavior of the group. Another, perhaps more common, term for credal uniqueness is making decision under *risk* (Luce and Raiffa, 1957; Einhorn and Hogarth, 1986). Levi (1980) relaxes the restriction of only one allowable probability model to admit a notion of "credal convexity," meaning that the set of admissible probability models is a convex set. We shall develop our theory under the assumption of credal uniqueness.[1] To pursue the convex case here, however, would distract from our present aim to introduce an alternative way to deal with randomness with respect to conventional approaches. Consequently, we will restrict our attention to the credally unique case.

5.1 Unification

The classical approach to computing expected utility is to form the product of the utility of each possible outcome and the probability of that outcome being realized, and then to sum these products over all outcomes. This approach is completely general in that it does not require any special syntactic relationships to exist between the utilities and the probabilities. By exploiting the close syntactic relationship between utilities and probabilities as developed herein, however, it becomes possible to merge the two component into a seamless characterization of the preferential and probabilistic aspects of a multistakeholder decision problem.

To extend to environments where uncertainty exists, we follow the approach introduced by Howard and Matheson (1984) and applied by Keeney and Raiffa (1993), Russell and Norvig (2003), Shoham and Leyton-Brown (2009), and others, who view a multistakeholder decision problem as an *influence diagram* or *decision network* comprising *decision nodes*, which correspond to decisions that are entirely under the control of the stakeholders, and *chance nodes*, which correspond to decisions that are under the control of random phenomena. These

---

[1] The reader interested in the credally convex case is referred to Levi (1980), especially chapters 5 and 6.

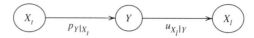

Figure 5.1. The social influence diagram for the party example.

diagrams involve multiple kinds of nodes, and thus they are not Bayesian networks, because, although the edges between decision nodes and chance nodes
represent influence, the influence is not characterized by probability mass functions. Thus, although the diagrams illustrate useful visual relationships, the
edges cannot be interpreted in the same way that the edges of a Bayesian
network can be used; namely, as elements of a multivariate probability mass
function.

An important consequence of the way utilities are defined in Chapter 2, however, eliminates the syntactical distinction between decision nodes and chance
nodes and permits a unified mathematical development of the deterministic
and random components of a multistakeholder decision problem. To motivate
a unified approach, let us consider a simple example.

Example 5.1 Suppose you want to hold a party, either Friday (f) or Saturday
(s). The location will be either an indoor party (i) or an outdoor party (o). The
weather on either day will either be rainy (r) or dry (d). Thus, there are eight
distinct outcomes:

$$\{(f,i,r),(f,i,d),(f,o,r),(f,o,d),(s,i,r),(s,i,d)(s,o,r),(s,o,d)\}.$$

This problem has deterministic components (the time and location) and a
stochastic component (the weather state). Although there are many ways to
frame this problem, one logical way is to consider the influence diagram illustrated in Figure 5.1, where the time node X_t influences the weather node Y,
which in turn influences the location node X_l. The edges of this diagram may
be defined as follows. Since the probability of the weather state depends only
on the time the party is held, we may express this relationship by the directed
edge $(X_t \rightarrow Y)$. We may represent this edge mathematically by the probability
mass function for the weather state conditioned on the time state, denoted $p_{Y|X_t}$.
Also, the influence of the weather state on the location state is represented by
the directed edge $(Y \rightarrow X_l)$, which is mathematically expressed by the utility
of the location state conditioned on the weather state, denoted $u_{X_l|Y}$.

An important thing to appreciate about this formulation is that both $p_{Y|X_t}$
and $u_{X_l|Y}$ are mass functions. The distinction between them is that $p_{Y|X_t}$ is an
epistemological quantity that conditionally orders the degree of belief regarding
weather conditions, given the time state, and $u_{X_l|Y}$ is a praxeological quantity

that conditionally orders the preferences for the location, given the weather state. The difference between these two ordering functions, however, is purely semantic; since both are mass functions, they have the same syntax. By applying the chain rule, we may define a joint utility-probability mass function that characterizes the joint relationship of the time, weather, and location states as the product

$$u_{X_t Y X_l}(a_t, b, a_l) = u_{X_t}(a_t) p_{Y|X_t}(b|a_t) u_{X_l|Y}(a_l|b),$$

where $a_t \in \{f, s\}$, $b \in \{r, d\}$, $a_l \in \{i, o\}$, and u_{X_t} is X_t's categorical utility that orders the preferences over the time states.

Given this joint formulation of the epistemological and praxeological components of a decision problem, we may compute the marginal mass function for the time-location states by summing this joint utility-probability mass function over the weather states, resulting in the expected utility

$$u_{X_t X_l}(a_t, a_l) = \sum_b u_{X_t Y X_l}(a_t, b, a_l) = \sum_b u_{X_t}(a_t) p_{Y|X_t}(b|a_t) u_{X_l|Y}(a_l|b).$$

$\square$

Since the utilities and probabilities in Example 5.1 all possess the same syntactic structure (namely, they are mass functions), there is no mathematical reason for not forming a joint mass function by invoking the chain rule. If the context were changed so that the functions u_{X_t} and $u_{X_l|Y}$ defined in Example 5.1 were probabilities rather than utilities, there would be no question about applying the chain rule. But, given the application in question, we are intermixing praxeological and epistemological quantities in a seemingly unconventional way. It is essential to remember, as discussed in Appendix A, that theory does not dictate or restrict application. Certainly, there can be no deep philosophical impediments to interweaving the praxeological and epistemological influence relationships. At the end of the day, the mathematics of probability is nothing more than a tool, and, as with any tool, its successful use depends on the skill of the one wielding it.

We now formalize our development by introducing notation to account for the deterministic and stochastic components of a problem. As before, we let $\mathcal{X}_n = \{X_1, \ldots, X_n\}$ denote a group of n stakeholders, and we assume that each X_i is able to make a deterministic choice a_i from its action set $\mathcal{A}_i$.[2] In addition to this set of deterministic stakeholders, we must also consider a set of stochastic

---

[2] It is possible to generalize this problem further by introducing so-called randomized decision rules, such that the X_i's choose an action according to a given probability distribution. Such a generalization, while perhaps of theoretical interest, would introduce a needless complication in our development, hence will not be pursued.

stakeholders, $\mathcal{Y}_m = \{Y_1, \ldots, Y_m\}$. These stakeholders are also capable of making decisions, but those decisions are random; they are unknown to all X_i, and no X_i has control over them. The randomness can have several sources, such as the unknown state of nature, the unknown actions of other stakeholders, and so forth. We assume that each Y_j can choose from a stochastic action set $\mathcal{B}_j$. Let $\mathcal{B} = \mathcal{B}_1 \times \cdots \times \mathcal{B}_m$ denote the joint stochastic action space. We may combine these two categories of stakeholders into the combined group $\mathcal{X}_n \cup \mathcal{Y}_m = \{X_1, \ldots, X_n, Y_1, \ldots, Y_m\}$.

Given this combined stakeholder group, we must extend the definition of a joint conjecture as introduced in Definition 2.2 to account for stochastic elements. We must also extend the notion of a concordant utility as introduced in Definition 2.11.

Definition 5.1   Let $\{X_1, \ldots, X_n, Y_1, \ldots, Y_m\}$ be a combined deterministic and stochastic stakeholder group with product action space $\mathcal{A} \times \mathcal{B}$. A *stochastic joint conjecture* is a profile $\{\mathbf{a}_1, \ldots, \mathbf{a}_n, b_1, \ldots, b_m\}$, where $\{\mathbf{a}_1, \ldots, \mathbf{a}_n\}$ is a joint conjecture for $\{X_1, \ldots, X_n\}$ and $\{b_1, \ldots, b_m\}$ is the antecedent of a hypothetical proposition that Y_j will instantiate b_j, $j = 1, \ldots, m$.   □

Definition 5.2   Let $\{X_1, \ldots, X_n, Y_1, \ldots, Y_m\}$ be a combined deterministic and stochastic stakeholder group with product action space $\mathcal{A} \times \mathcal{B}$ and let $\{\mathbf{a}_1, \ldots, \mathbf{a}_n, b_1, \ldots, b_m\}$ be a stochastic joint conjecture. A *stochastic concordant utility* is a function $U_{X_1 \cdots X_n Y_1 \cdots Y_m} \colon \mathcal{A}^n \times \mathcal{B} \to [0, 1]$ that characterizes all of the social relationships that exist among the deterministic stakeholders, the relationships that exist among the stochastic stakeholders, the relationships that the deterministic stakeholders exert on the stochastic stakeholders, and the relationships that the stochastic stakeholders exert on the deterministic stakeholders.   □

As developed in Chapter 2, a concordant utility is constructed by the aggregation of individual utilities by the application of (2.17). Since we wish to extend this procedure to the stochastic case, we must impose restrictions on the relationships between stakeholders to ensure that the aggregation is well defined. These restrictions are most easily understood via graph theory.

We may create a network by specifying all of the deterministic and stochastic stakeholders as nodes and placing directed edges between them when social influence exists. To ensure that the resulting graph possesses the desired mathematical properties, we must insist upon a condition of acyclicity, such that no cycles occur in the influence relationships that exist among the stakeholders.

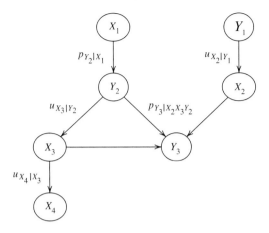

Figure 5.2. A utility-probability network.

Definition 5.3 A *utility-probability network* is a directed acyclic graph whose nodes consist of both deterministic and stochastic stakeholders, and whose edges are conditional mass functions that define the social and probabilistic influence relationships among the stakeholders. □

Figure 5.2 provides an illustration of a utility-probability network. Notice that this network interweaves the relationships among the deterministic and stochastic stakeholders such that deterministic stakeholders influence both deterministic and stochastic stakeholders, and stochastic stakeholders influence both deterministic and stochastic stakeholders.

Assuming that a stochastic multistakeholder decision problem can be expressed as a utility-probability network, the stochastic concordant utility can be constructed as follows: Suppose X_i has p_i deterministic stakeholder parents and q_i stochastic stakeholder parents, and Y_j has r_j deterministic stakeholder parents and s_j stochastic stakeholder parents. Define

$$\mathrm{pa}\,(X_i) = \{X_{i_1}, \ldots, X_{i_{p_i}}, Y_{k_1}, \ldots, Y_{k_{q_i}}\}$$

$$\mathrm{pa}\,(Y_j) = \{X_{j_1}, \ldots, X_{j_{r_j}}, Y_{l_1}, \ldots, Y_{l_{s_j}}\}.$$

For any stochastic joint conjecture $\{\mathbf{a}_1, \ldots, \mathbf{a}_n, b_1, \ldots, b_m\}$, let $\{\mathbf{a}_{i_1}, \ldots, \mathbf{a}_{i_{p_i}}, b_{k_1}, \ldots, b_{k_{q_i}}\}$ and $\{\mathbf{a}_{j_1}, \ldots, \mathbf{a}_{j_{r_j}}, b_{l_1}, \ldots, b_{l_{s_j}}\}$ denote the stochastic joint conjectures of $\mathrm{pa}\,(X_i)$ and $\mathrm{pa}\,(Y_j)$, respectively. The stochastic concordant utility is of the form

$$U_{X_1\cdots X_n Y_1\cdots Y_m}(\mathbf{a}_1,\ldots,\mathbf{a}_n,b_1,\ldots,b_m) =$$

$$\prod_{i=1}^{n} u_{X_i|\text{pa}(X_i)}(\mathbf{a}_i|\mathbf{a}_{i_1},\ldots,\mathbf{a}_{i_{p_i}},b_{k_1},\ldots,b_{k_{q_i}})$$

$$\prod_{j=1}^{m} p_{Y_j|\text{pa}(Y_j)}(b_j|\mathbf{a}_{j_1},\ldots,\mathbf{a}_{j_{r_j}},b_{l_1},\ldots,b_{l_{s_j}}).$$

Once the concordant utility is defined, the next step is to compute the expected concordant utility as the marginal concordant utility obtained by summing over the stochastic states, yielding the *expected concordance utility*

$$U_{X_1\cdots X_n}(\mathbf{a}_1,\ldots,\mathbf{a}_n) = \sum_{b_1,\ldots,b_m} U_{X_1\cdots X_n Y_1\cdots Y_m}(\mathbf{a}_1,\ldots,\mathbf{a}_n,b_1,\ldots,b_m).$$

Once the expected concordant utility is available, the solution concepts defined in Chapter 3 can be applied.

It is of interest to notice that we may also compute the joint probability mass function of the stochastic subgroup $\{Y_1,\ldots,Y_m\}$ as the stochastic marginal

$$p_{Y_1\cdots Y_m}(b_1,\ldots,b_n) = \sum_{\mathbf{a}_1,\ldots,a_n} U_{X_1\cdots X_n Y_1\cdots Y_m}(\mathbf{a}_1,\ldots,\mathbf{a}_n,b_1,\ldots,b_m).$$

To illustrate, consider the utility-probability network displayed in Figure 5.2. The corresponding stochastic concordant utility is

$$U_{X_1 X_2 X_3 X_4 Y_1 Y_2 Y_3}(\mathbf{a}_1,\mathbf{a}_2,\mathbf{a}_3,\mathbf{a}_4,b_1,b_2,b_3) =$$

$$u_{X_1}(\mathbf{a}_1)u_{X_2|Y_1}(\mathbf{a}_2|b_1)u_{X_3|Y_2}(\mathbf{a}_3|b_2)u_{X_4|X_3}(\mathbf{a}_4|\mathbf{a}_3)$$

$$p_{Y_1}(b_1)p_{Y_2|X_1}(b_2|\mathbf{a}_1)p_{Y_3|X_3 X_3 Y_2}(b_3|\mathbf{a}_2,\mathbf{a}_3,b_2),$$

and the expected concordant utility is the deterministic marginal

$$U_{X_1 X_2 X_3 X_4}(\mathbf{a}_1,\mathbf{a}_2,\mathbf{a}_3,\mathbf{a}_4) = \sum_{b_1,b_2,b_3} U_{X_1 X_2 X_3 X_4 Y_1 Y_2 Y_3}(\mathbf{a}_1,\mathbf{a}_2,\mathbf{a}_3,\mathbf{a}_4,b_1,b_2,b_3).$$

Example 5.2 Let us now expand the manufacturing problem discussed in Example 3.2 to include random components. A company is to build a factory in one of two different locations $\{L_1, L_2\}$ to market one of four different widgets, $\{W_1, W_2, W_3, W_4\}$, each of which can be manufactured by any of four different process, $\{P_1, P_2, P_3, P_4\}$. The problem is to choose from among the 32 possible alternatives $\mathbf{a}_{ijk} = (L_i, P_j, W_k)$, where $i \in \{1,2\}$ and $j,k \in \{1,2,3,4\}$. The preference for location is based on such factors as the availability of an appropriate labor force, transportation costs, and other social and economic factors; the preference for the manufacturing process is based on the impact to the environment, which in turn is influenced by the location; and the preference for the product is based on its profitability.

We must also take into consideration the ecological status of each location with respect to the environmental impact and the economic climate regarding the profitability of the product they will manufacture. The ecological conditions, such as long-term projections of population density, long-term climate conditions, and so forth, are random phenomena, as are the economic conditions that can affect profitability. We assume there are two distinct ecological environments $\{V_1, V_2\}$ and two distinct economic environments $\{E_1, E_2\}$. Thus, the choice of alternatives must be based on the expected values with respect to these random phenomena.

This multistakeholder decision problem involves five attributes (three are deterministic stakeholders and two are stochastic stakeholders) defined as follows:

$$X_1 = \text{Location of factory}$$

$$X_2 = \text{Manufacturing process}$$

$$X_3 = \text{Product to be manufactured}$$

$$Y_1 = \text{Ecological environment}$$

$$Y_2 = \text{Economic environment}.$$

Since each location results in its own ecological environment, the environmental impact of each manufacturing process is influenced by these conditions. Also, the profitability of each process is affected by the choice of manufacturing process and by the economic environment. Figure 5.3 illustrates the utility-probability network corresponding to the influence flows among the attributes. The mass function $p_{Y_1|X_1}$ is the probability of the ecological environment as conditioned on the location; $u_{X_2|Y_1}$ is the conditional utility of the manufacturing process given the environmental state; and $u_{X_3|X_2Y_2}$ is the conditional utility of the profit on both the manufacturing process and the economic conditions. The stochastic concordance utility associated with this decision problem is

$$U_{X_1X_2X_3Y_1Y_2}(a_1,a_2,a_3,b_1,b_2) = u_{X_1}(a_1)u_{X_2|Y_1}(a_2|b_1)$$

$$u_{X_3|X_2Y_2}(a_3|a_2,b_2)$$

$$p_{Y_1|X_1}(b_1|a_1)p_{Y_2}(b_2).$$

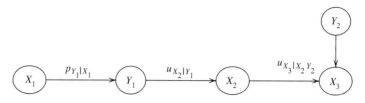

Figure 5.3. The utility-probability network for the manufacturing scenario.

We are given two location alternatives for X_1, four process alternatives for X_2, and four product choices for X_3. Thus, the action spaces are

$$\mathcal{A}_1 = \{L_1, L_2\}$$
$$\mathcal{A}_2 = \{P_1, P_2, P_3, P_4\}$$
$$\mathcal{A}_3 = \{W_1, W_2, W_3, W_4\}$$

and the product action space is $\mathcal{A} = \mathcal{A}_1 \times \mathcal{A}_2 \times \mathcal{A}_3$, consisting of all ordered triples $\mathbf{a} = (a_1, a_2, a_3)$, where $a_i \in \mathcal{A}_i$. Also, the stochastic stakeholder realization spaces are

$$\mathcal{B}_1 = \{V_1, V_2\}$$
$$\mathcal{B}_2 = \{E_1, E_2\}$$

and the product realization space is $\mathcal{B} = \mathcal{B}_1 \times \mathcal{B}_2$, consisting of all ordered pairs $\mathbf{b} = (b_1, b_2)$.

To proceed with this example, we must supply numerical values for the utilities associated with each of the states. Starting with the location state, let us define the categorical utility for X_1 as

$$u_{X_1}(a_1) = \begin{cases} 0.4 & \text{if } a_1 = L_1 \\ 0.6 & \text{if } a_1 = L_2 \end{cases} . \tag{5.1}$$

Since a different ecological environment corresponds to each location, the probability mass function associated with the environmental impact is conditioned on the location, as defined in Table 5.1.

Since the environmental impact of the manufacturing processes depends on the ecological environment, the corresponding utility is conditional, as illustrated in Table 5.2.

Table 5.3 illustrates the profit utility conditioned on the manufacturing process and the economic environment.

Table 5.1. *Conditional*
ecological probability
$p_{Y_1|X_1}(V_l|L_i)$

	L_1	L_2
V_1	0.35	0.55
V_2	0.65	0.45

Table 5.2. *Environmental impact*
utility $u_{X_2|Y_1}(P_j|V_l)$

	V_1	V_2
P_1	0.3571	0.1250
P_2	0.2857	0.3750
P_3	0.2143	0.3325
P_4	0.1429	0.1675

Table 5.3. *Profit conditional utility* $u_{X_3|X_2Y_2}(W_k|P_j,E_m)$ *for each manufacturing*
process and economic environment

	P_1,E_1	P_2,E_1	P_3,E_1	P_4,E_1		P_1,E_2	P_2,E_2	P_3,E_2	P_4,E_2
W_1	0.1667	0.0870	0.1852	0.0909	W_1	0.1782	0.2557	0.0995	0.4443
W_2	0.1666	0.3043	0.3333	0.2727	W_2	0.4523	0.1320	0.4328	0.3459
W_3	0.2500	0.4348	0.4444	0.1818	W_3	0.1450	0.3250	0.3554	0.1132
W_4	0.4167	0.1739	0.0370	0.4545	W_4	0.2245	0.2873	0.1131	0.0966

Finally, the economic environment probability is given by

$$p_{Y_2}(b_2) = \begin{cases} 0.345 & \text{if } b_2 = E_1 \\ 0.655 & \text{if } b_2 = E_2 \end{cases}.$$

Since this a completely dissociated scenario, the group welfare function
coincides with the concordant utility and is given by

$$v_{X_1X_2X_3Y_1Y_2}(a_1,a_2,a_3,b_1,b_2) = u_{X_1}(a_1)u_{X_2|Y_1}(a_2|b_1)$$
$$u_{X_3|X_2Y_2}(a_3|a_2,b_2)$$
$$p_{Y_1|X_1}(b_1|x_1)p_{Y_2}(b_2).$$

Table 5.4. *Expected group welfare function* $v_{X_1X_2X_3}(L_i, P_j, W_k)$ *for the manufacturing problem.*

	L_1, P_1	L_1, P_2	L_1, P_3	L_1, P_4		L_2, P_1	L_2, P_2	L_2, P_3	L_2, P_4
W_1	0.0144	0.0240	0.0203	0.0111	W_1	0.0299	0.0386	0.0317	0.0182
W_2	0.0292	0.0486	0.0412	0.0225	W_2	0.0290	0.0374	0.0307	0.0177
W_3	0.0150	0.0249	0.0211	0.0115	W_3	0.0550	0.0710	0.0582	0.0335
W_4	0.0240	0.0400	0.0339	0.0185	W_4	0.0376	0.0485	0.0398	0.0229

The expected group welfare function is obtained by summing over the stochastic states, yielding

$$v_{X_1X_2X_3}(a_1, a_2, a_3) = \sum_{b_1, b_2} v_{X_1X_2X_3Y_1Y_2}(a_1, a_2, a_3, b_1, b_2),$$

with values displayed in Table 5.4.

Straightforward calculations yield the conditioned nondominated set

$$N = \{(L_1, P_2, W_3), (L_1, P_3, W_2), (L_2, P_2, W_3), (L_2, P_3, W_2)\},$$

with the globally maximizing alternative

$$\mathbf{a}^* = \arg \max_{a_1, a_2, a_3} \{v_{X_1X_2X_3}(a_1, a_2, a_3)\} = (L_2, P_2, W_3). \tag{5.2}$$

The expected individual welfare functions are obtained as

$$v_{X_i}(a_i) = \sum_{\sim a_i} v_{X_1X_2X_3}(a_1, a_2, a_3).$$

Since X_1 possesses a categorical utility, its ex post expected welfare function is the same as (5.1). The ex post expected welfare functions for X_2 and X_3 are displayed in Tables 5.5 and 5.6, respectively. By inspection, we see that the negotiating rectangle is

$$R_{X_1X_2X_3} = \{L_2, P_2, W_3\},$$

which corresponds, in this case, with the globally maximizing alternative. $\square$

Table 5.5. *Expected welfare function v_{X_2} for the manufacturing problem*

P_1	P_2	P_3	P_4
0.234	0.333	0.277	0.156

Table 5.6. *Expected welfare function v_{X_3} for the manufacturing problem*

W_1	W_2	W_3	W_4
0.188	0.256	0.290	0.265

5.2 Summary

This chapter combines two notions of uncertainty into a single mathematical structure. Praxeological uncertainty, as manifest through opportunity cost, characterizes the behavioral ambivalence associated with making choices among alternatives, and epistemological uncertainty, as manifest through randomness, characterizes the ambivalence with respect to belief regarding the state of nature. This unification permits a departure from the conventional way of computing expected utility. With the classical formulation, utilities and probabilities have different mathematical structures, thus the expected utility operation is obtained by computing the sum of the products of the utilities and the corresponding probabilities. With our unified approach, both preferential and probabilistic aspects of the problem can be expressed with the same mathematical syntax, namely, conditional mass functions. Expected utility is obtained as the marginal utility of the deterministic stakeholders.

This approach also provides a natural symmetry between the preferential model, as defined by the utilities, and the probabilistic model, as defined by the probability mass functions. This symmetry permits the utilities for the stakeholders to be conditioned on the values that the stochastic stakeholders assume, and it also permits the mass functions of the stochastic stakeholders to be conditioned on the conjectures of the stakeholders. No such natural symmetry exists with the classical approach, since all utilities are categorical and do not necessarily possess the structure of mass functions. Consequently, although the deterministic stakeholder utilities are functions of stochastic stakeholders' actions, they are not conditioned, in the probabilistic sense, on their actions.

6

Satisficing

Two souls, alas, do dwell within his breast; The one is ever parting from the other.

— Goethe
Faust, Part I

Optimization is the sine qua non of decision theory. The list of contributors to the concept is a veritable who's who of mathematics that includes Fermat, Newton, Gauss, Euler, Lagrange, Fourier, Edgeworth, Pareto, von Neumann, Wiener, Dantzig, Kantorovich, Bellman, Kalman, Arrow, Nash, and others. Indeed, optimization has played a central role in virtually every decision-making procedure and enjoys uncontested mathematical respectability. As Euler noted, "Since the fabric of the world is the most perfect and was established by the wisest Creator, nothing happens in this world in which some reason of maximum or minimum would not come to light" (cited in Polya, 1954).

With the exception of work based on the results of Edgeworth, Pareto, von Neumann, Arrow, and Nash, however, optimization theory has focused on the behavior of a single decision maker. Indeed, the concept of optimization is an individual concept. In group scenarios, the issues become more complex: If a group wishes to optimize, it must act as if it were a single entity. As Arrow's (1951) impossibility theorem establishes, however, it is not generally possible to define a preference ordering for a group in terms of the preference orderings of its individual members. Consequently, the concept of optimization in group settings is often expressed through such concepts as equilibrium, nondominance, and social welfare, none of which enjoy the type of global superlativeness that the term *optimization* typically connotes. Hence, the notion of optimization loses much of its significance in group settings. Nevertheless, much effort has been devoted to the development of ways to recast multiple-stakeholder decision problems under an optimization umbrella.

There are at least three conventional approach to forming group decisions that have received lots of attention. The first we shall discuss is to define some notion of optimization that takes into account the conflicts that exist among the stakeholders. Several possibilities exist when a separate utility exists for each individual or criterion. One approach is to aggregate the utilities to form a global utility for the group, thereby transforming a multicriterion or multiagent problem into a single-criterion or single-agent (superplayer) problem for which optimization can be applied. The most common form of aggregation is to form a weighted sum of the utilities, where the weights essentially perform the function of converting all of the utilities to a common scale. Then different choices for the weights result in different notions of what is optimal.

A second approach is to keep the utilities separate and form the Pareto frontier of nondominated consequences. Then each member of this set corresponds to a solution that can be interpreted as optimal in the sense that the benefit to any one stakeholder cannot be improved without decreasing the benefit to at least one other stakeholder.

A third approach is to invoke game theory and apply a solution concept such as Nash equilibria that corresponds to a jointly constrained optimal solution such that if any individual were to change its decision, its utility would decrease. Nash solutions, however, can be very conservative and result in mediocre performance. For example, consider the Prisoner's Dilemma, where the Nash solution is worse for both players than is the Pareto solution.

What these different approaches have in common is that they all seek to recast an intrinsically multifaceted problem as a single-faceted one in order to apply the principle of optimization. There are at least two major motivations for such a procedure. First, the procedure transforms a complex problem into a comparatively simple problem. Second, it permits the decision maker to apply well-known and trusted solution techniques. Zadeh (1958) counseled a half-century ago, however, against relying too heavily on the concept of optimization to justify a decision.

> Not too long ago we were content with designing systems which merely met given specifications. . . . Today, we tend, perhaps, to make a fetish of optimality. If a system is not the "best" in one sense or another, we do not feel satisfied. Indeed, we are apt to place too much confidence in a system that is, in effect, optimal by definition. At present, no completely satisfactory rule for selecting decision functions is available, and it is not very likely that one will be found in the foreseeable future. Perhaps all that we can reasonably expect is a rule which, in a somewhat equivocal manner, would delimit a set of "good" designs for a system. (Zadeh, 1958, p. 3)

The most defensible motive for relying on optimization is to exploit all of the useful structure and information contained in the model used to describe the decision problem – any other solution leaves something on the table. The problem in multistakeholder contexts, however, is that what is optimal for one stakeholder is usually not optimal for another, and someone must leave something on table. In light of this situation, let us consider employing a notion of rationality that is softer and more flexible than optimization. This is not a new idea. As Arrow (1986) observed, "Among the classical economists, such as Smith and Ricardo, rationality had the limited meaning of preferring more to less" (p. 204). Furthermore, Levi (1997) argues that "the fixed principles of coherent or consistent choice, belief, desire, etc. will have to be weak enough to accommodate a wide spectrum of potential changes in point of view" (p. 24). In light of these observations, we suggest that always insisting on optimization may be shooting beyond the mark when the interests of multiple stakeholders are involved. A more robust and flexible approach is to back away from the demand for optimization and settle for an approach that can be rigorously justified as being "good enough."

It is a platitude that decision makers should make the best choices possible, but they cannot rationally choose an alternative, even if they do not know of anything better, unless they know that it is good enough. Being good enough is the fundamental desideratum of rational decision makers – being best is a bonus. For this desideratum to be useful, however, we must precisely define what it means to be good enough. Mathematically formalizing such a concept is not as straightforward as optimizing. Being best is an absolute concept but being good enough is not an absolute and can be expressed in degrees. Consequently, we must not demand or expect a unique good-enough solution; instead, we must be willing to consider multiple solutions with varying degrees of adequacy.

Optimization is a very sophisticated concept and requires that the decision maker rank the alternatives according to its preferences for the consequences of implementing them. Although conventional utility theory maps alternatives into numerical degrees of satisfaction, utilities must also account for any dissatisfaction that may accrue to the decision maker. Conventional utility theory aggregates these two polarizing aspects into a single function that characterizes the net utility of the alternative. A well-known example in an engineering context is the performance index, or cost function,[1] to be minimized when designing an optimal regulator. The performance index consists of a component involving the distance of the actual state from the desired state and a component involving the energy required to effect the control. Essentially, the control engineer's task

---

[1] It is often more convenient to define cost as the negative of utility.

is to achieve an acceptable balance between performance and cost by adjusting the relative weights of the two components of the performance index. Thus, even though only one function is minimized, two issues must be considered.

As a practical matter, however, people often separate the polarizing aspects of a decision problem into two components. Attached to virtually every nontrivial option are attributes that are desirable and attributes that are not desirable. For example, to increase performance, one usually expects to pay more. To win a larger reward, one expects to take a greater risk. People naturally evaluate the upside versus the downside, the pros versus the cons, the pluses versus the minuses, the benefits versus the costs. One simply evaluates tradeoffs option by option – putting the gains and the losses on the balance to see which way it tips. The result of evaluating dichotomies in this way is that the benefits of taking the action must be at least as great as the costs. In this sense, such evaluations provide a distinct notion of being good enough.

Let us take the notion of "getting at least what you pay for" as an operational definition of being good enough. This notion of rational behavior may be a natural concept for people, but, perhaps not surprisingly, it has received far less attention in the literature as a basis for formalized decision making than the optimality paradigm, which is the philosophy that only the very best will do. The optimization paradigm is usually taken for granted as the standard against which all approaches should be measured, but people may not be that rigid in their demands for strict optimality (see Slote, 1989). Furthermore, algorithms that are designed to search for the optimum cannot evaluate options that are not the best. This situation may be likened to climbing a mountain in the fog. A climber can know he is at the summit only when he stops ascending, but he cannot know how close he is to the top before he achieves that goal.

For this notion of rationality to be useful, we must establish a rigorous mathematical framework that covers both single-stakeholder and multistakeholder applications. To proceed, we must first define an operational concept to serve as an alternative to optimization, while at the same time retaining an unambiguous mathematical description of the notion. Once that concept has been established, we must provide a rigorous mathematical definition of what it means to be good enough. Following the pattern established in previous chapters, we motivate our approach by drawing a praxeological analog to a familiar epistemological concept which, in this context, is the avoidance of error.

6.1 Error avoidance

It ain't so much ignorance that ails mankind as it is knowing so much that ain't so.

— *Josh Billings*

A key to the successful implementation of the praxeological coherence principle is the operational concept of *acceptability*. If every stakeholder insists that what is best for itself must be seriously considered by the group as acceptable, and if the group were to equate acceptability with optimality, then an impasse cannot generally be avoided, since it is likely that at least some of the stakeholders will not achieve their optimal result. The only way to overcome this problem is for the individual stakeholders and the group all to function in accordance with an operational definition of acceptability that is softer and more flexible than the rigid and brittle notion of optimization. To motivate such a new concept of acceptability, let us consider an observation by James (1956).

> There are two ways of looking at our duty in the matter of opinion – ways entirely different, and yet ways about whose difference the theory of knowledge seems hitherto to have shown very little concern. We must know the truth, and we must avoid error – these are our first and great commandments as would-be knowers; but they are not two ways of stating an identical commandment, they are two separable laws...
>
> Believe truth! Shun error! – these, we see, are two materially different laws; and by choosing between them we may end by coloring differently our whole intellectual life. We may regard the chase for truth as paramount, and the avoidance of error as secondary; or we may, on the other hand, treat the avoidance of error as more imperative, and let truth take its chance. (pp. 17, 18)

Those who are committed to seeking the truth and nothing but the truth may find themselves without sufficient information to determine what is true with enough confidence to impel action. On the other hand, those who are committed to avoiding error would be willing to weigh the evidence and focus attention on propositions for which the evidence is sufficient either to retain them as being likely to be true or to reject them from serious consideration because they are likely to be false. The difference between these two mindsets is the degree of caution exercised in making a choice. Those who are committed to avoiding error will find it easier to suspend judgment when the evidence is not conclusive, while those who are committed to the truth and nothing but the truth may be so anxious to obtain an answer that they cannot bring themselves to suspend judgment.

Example 6.1 Suppose you are hiking in the Himalayas and you see in the distance a large, hairy two-legged creature. To explain what you see, four propositions come to mind:

$$\omega_1 = \text{a hallucination because of a lack of oxygen}$$
$$\omega_2 = \text{a large, scruffy human being}$$

ω_3 = an actor dressed as an abominable snowman
ω_4 = a real yeti

Suppose, using your training as a decision theorist, you are willing to assign utilities and probabilities to each of these propositions, compute the expected utility of each, and take an action corresponding to the proposition with the highest expected utility. If you behave in this way, you are acting in accordance with a "truth seeker" mentality in the sense described by James (1956). Deciding what to believe is your primary objective – not making an error is of secondary importance.

On the other hand, suppose you are, in the Jamesian sense, an "error-avoider." If so, you would be reluctant to focus on a single proposition unless you had overwhelming evidence confirming its truth. Rather than deciding which single proposition you should accept, you might consider which, if any, of the propositions are not worthy of retention. If you were not willing to risk any error at all, you would not reject any of the propositions, but you would also have made a vacuous decision – a condition of paralysis. Presumably, however, the fact that you have experienced some phenomenon means that you should respond in some way, such as taking a photograph, protecting yourself, or quickly seeking a lower altitude to keep from passing out because of a lack of oxygen.

If no response is desired or needed, then you would be justified in the interest of avoiding error for not eliminating any of the propositions – to make no decision. But if a response is needed, then you might ask yourself: "Is the information I gain by eliminating one or more propositions worth the risk of making an error?" This question is fundamentally different from one a truth-seeker would ask.   □

An error-avoider enjoys a somewhat subtle, if temporary, advantage over a truth-seeker. Truth-seeking is an absolute concept – it does not come in degrees – and a truth-seeker is obligated to settle on one and only one proposition. On the other hand, error-avoiding is not an absolute; there are various degrees of error avoidance, depending on the amount and quality of information available. If there is sufficient information available, one could eliminate all but one proposition – an indirect way of "backing in" to the truth. Otherwise, there may be several noneliminated propositions. The net result is that an error-avoider may refine its choices to the extent warranted by the information, but is not required to select, as if by fiat, a unique "best" choice. This approach has its limitations, of course, since taking action requires the decision maker to settle ultimately on a unique choice. Nevertheless, much can be gained by viewing decision making from the perspective of an error-avoider.

6.1.1 Theoretical background

It is more important that a proposition be interesting than that it be true.
This statement is almost a tautology.

— *Alfred North Whitehead*
Adventures of Ideas
(Cambridge University Press, 1933)

Levi (1980) has defined a mathematically precise approach to implement the error-avoidance framing of a decision problem. Although the proximate aim of the decision maker is to avoid error, this aim is tempered by the demand for information. Information, in this context, is determined by the degree of refinement obtained by eliminating propositions from serious consideration. Levi constructs two utilities to account for these competing criteria: an error avoidance utility and an informational value utility.

Let $\Omega = \{\omega_1, \omega_2, \ldots, \omega_n\}$ be a finite set of propositions, one and only one of which is true, and let $\mathcal{F}$ denote a Boolean algebra[2] of subsets of Ω. Then for any set $B \in \mathcal{F}$, we say that B is true if and only if B contains the true element. We may then define the error avoidance utility as

$$T(B) = \begin{cases} 1 & \text{if } B \text{ is true} \\ 0 & \text{otherwise} \end{cases}.$$

We may interpret B as the set of propositions that will not be eliminated. This utility illustrates a major distinction between truth-seeking and error-avoiding: a truth-seeking utility would require B to be a singleton set, while an error-avoiding framing relaxes that constraint. It is straightforward to see that $T(\varnothing) = 0$ and that T is additive over disjoint sets; that is, $T(B_1 \cup B_2) = T(B_1) + T(B_2)$ if $B_1 \cap B_2 = \varnothing$. Thus, T is a normalized measure over $\mathcal{F}$.

Our desire is to find the smallest element of $\mathcal{F}$ that is consistent with the demand for information while avoiding error. The only way a decision maker can be completely sure of avoiding error is to set $B = \Omega$, thereby eliminating no propositions. Doing so, however, would result in a vacuous decision problem that would deny the decision maker any improvement in information. To temper the desire to avoid error with the demand for information, we must compute the informational value of each B. For reasons that will subsequently become clear, we will determine this value by first considering the informational value of rejecting B.

---

[2] A Boolean algebra of sets of Ω is a collection of subsets of Ω that contains $\varnothing$ and Ω, and is closed under complementation, finite union, and finite intersection.

Independently of whether or not it is true, B is of some intrinsic informational value to the decision maker. Such valuations may take many forms, including economic, political, moral, cognitive, aesthetic, or personal value. In this regard, Popper (1963) observed that "we must also stress that *truth is not the only aim of science*. We want more than truth: what we look for is *interesting truth* [emphasis in original]" (p. 229). To define the utility of the informational value of rejection, it is reasonable that the following properties should hold. Let P_R be a utility function that quantifies the informational value of rejection. Following Levi (1980), we make the following assumptions regarding how informational value is computed.

1. Since X chooses sets rather than points, P_R should be a mapping over the Boolean algebra, $\mathcal{F}$.
2. Measures of informational value should be nonnegative and finite. Consequently, it is reasonable to suppose that there exists a unit of informational value; that is, the informational-value utility should range over the unit interval.
3. Rejecting none of the propositions is of no informational value, thus $P_R(\varnothing) = 0$. Also, rejecting all propositions is of maximal informational value, thus $P_R(\Omega) = 1$.
4. The informational value of rejecting any set of propositions that does not intersect any previously rejected set should be invariant to whatever else has been rejected.
5. The informational value of rejecting disjoint propositions sets must be additive. That is, if $B_1 \in \mathcal{F}$ and $B_2 \in \mathcal{F}$ and $B_1 \cap B_2 = \varnothing$, then rejecting both B_1 and B_2 is equivalent to rejecting $B_1 \cup B_2$; that is, $P_R(B_1 \cup B_2) = P_R(B_1) + P_R(B_2)$.

These conditions imply that P_R is a probability measure over $\mathcal{F}$. We emphasize that P_R differs from classical utilities in a very important way; namely, it is a mapping of sets of propositions – elements of the Boolean algebra $\mathcal{F}$ over the proposition space Ω, rather than a mapping of points in Ω – individual propositions. If we assume that the Boolean algebra contains all singleton sets, then we may define a utility $p_R(\omega) = P_R(\{\omega\})$. We observe that $p_R(\omega)$ is a mass function, that is, $p_R(\omega) \geq 0$ for all $\omega \in \Omega$ and $\sum_{\omega} p_R(\omega) = 1$. The measure and the mass function are related by

$$P_R(B) = \sum_{\omega \in B} p_R(\omega).$$

Although P_R is a probability measure, it does not possess any of the traditional semantic interpretations of probability. It is not a means of quantifying such attributes as belief, propensity, frequency, and so forth. Rather, it is used here solely as a means of quantifying informational attributes.

Levi (1980) terms P_R an *information-determining probability*.[3] This usage of the mathematics of probability is distinct from its traditional epistemological usage. In contrast to the usual interpretation of probability as a measure of the strength of an inductive inference, Levi uses the mathematical structure of probability theory to measure the strength of an *abductive* inference. An abductive inference is not an assertion of truth; rather, it is an assertion of the importance of a proposition, independent of its truth. Essentially, $P_R(B)$ may be viewed as the utility of the informational value that accrues to the decision maker if B is eliminated from consideration as a serious possibility. The motivation for defining informational value in this way is that we are developing the error-avoiding view of rejecting propositions rather than the truth-seeking view of accepting one and only one proposition. We interpret P_R as follows: Rejecting B means that the decision maker accrues $P_R(B)$ worth of informational value. Thus, the larger the set B, the more informational value is retained.

Once we have defined the informational value of rejection, we may compute the informational value of not rejecting B as $I(B) = 1 - P_R(B)$. We may now compute the utility of both avoiding error and acquiring information by not rejecting B as the convex combination of $P_R(B)$ and $I(B)$, yielding Levi's (1980) *epistemic utility function* $\phi(B) = \alpha T(B) + (1 - \alpha)I(B)$, where $\leq \alpha \leq 1$. The parameter α represents the relative importance that is attached to avoiding error versus acquiring information. Setting $\alpha = 1$ places a premium on avoiding error, and setting $\alpha = \frac{1}{2}$ places equal weight on the demand for information and the avoidance of error. Setting $\alpha < \frac{1}{2}$, however, could result in an erroneous answer being preferred to a correct answer. Thus, α should be restricted to the interval $[\frac{1}{2}, 1]$.

Since utilities are invariant to scale and zero level, we may transform ϕ by a positive affine transformation of the form

$$\phi^{\alpha}(B) = \frac{1}{\alpha}\phi(B) - \frac{1 - \alpha}{\alpha}$$

$$= T(B) - q P_R(B)$$

$$= \begin{cases} 1 - q P_R(B) & \text{if } B \text{ is true} \\ -q P_R(B) & \text{if } B \text{ is false} \end{cases},$$

where $q = \frac{1-\alpha}{\alpha}$. Thus, $0 \leq q \leq 1$.

---

[3] Information in this context is different from the usage of the term in Shannon information theory. Here, information is essentially synonymous with the communication of knowledge.

To complete the framework, we must consider the evidence regarding the propositions. Let P_S denote a probability measure over the Boolean algebra $\mathcal{F}$ such that $P_S(B)$ is the probability that B contains the true proposition. Thus, P_S is a measure of the strength of an inductive inference. In the parlance of epistemology, P_S is called a *credal probability*. Let p_S denote the mass function associated with P_S, such that

$$P_S(B) = \sum_{\omega \in B} p_S(\omega).$$

The probability measure P_S is an *expectation-determining* probability and is used to compute the expected value of epistemic utility, yielding

$$E\phi^\alpha(B) = [1 - q\,P_R(B)]P_S(B) - q\,P_R(B)[1 - P_S(B)]$$
$$= P_S(B) - q\,P_R(B)$$
$$= \sum_{\omega \in B}[p_S(\omega) - qp_R(\omega)].$$

We now see that expected epistemic utility is maximized by the largest element of $\mathcal{F}$ for which the probability of truth is at least as great as q times the informational value of rejection. Thus, expected epistemic utility is maximized by rejecting all and only those elements of $\mathcal{F}$ for which $p_S(\omega) < qp_R(\omega)$. Accordingly, let us denote the largest set of nonrejected propositions as

$$\Sigma_q = \{\omega \in \Omega : p_S(\omega) \geq qp_R(\omega)\}. \tag{6.1}$$

Following Levi (1980), we term Σ_q the set of *seriously possible propositions*. We may interpret q as a *boldness parameter*. As q increases, the decision maker rejects more propositions and becomes more willing to accept error in the interest of obtaining more information.

We may view (6.1) as an expression of optimality in the sense that it represents optimal error avoidance. Barring ties, a truth-seeker must choose a unique best proposition, say ω^*, for which the probability of error is

$$P_{error} = 1 - p_S(\omega^*). \tag{6.2}$$

However, the probability of error for an optimal error-avoider is

$$P_{error} = 1 - \sum_{\omega \in \Sigma_q} p_S(\omega). \tag{6.3}$$

The price to be paid for minimizing error is that the guarantee of a unique choice is sacrificed. One who analyzes propositions in this way is willing to trade informational value for error avoidance.

Example 6.2 Let us return to the Hiking-the-Himalayas scenario described in Example 6.1. It is essential to evaluate the informational importance of each proposition without taking into consideration the likelihood of it being true. Take, for example, proposition ω_1, that you are hallucinating because of lack of oxygen. That proposition would indicate that the thin air is getting to you – a life-threatening situation of considerable informational value, since you might pass out and die if you reject that proposition and do not quickly seek a lower altitude. Thus, the informational value that would accrue if you were to reject this hypothesis would be extremely low – say, $p_R(\omega_1) = 0.03$. Notice that this evaluation is determined completely independently of the likelihood that you are suffering from the lack of oxygen. Next, consider proposition ω_2, that what you see is just an unkempt man on his way to somewhere. That proposition might be of little value to you, and rejecting it would be of considerable informational value to you, for then you would concentrate your attention on more weighty propositions – say, $p_R(\omega_2) = 0.5$. Now consider proposition ω_3, that someone is trying to deceive you. That proposition would be of modest value to retain, with the main consequence to you of being embarrassed by being fooled by a practical joker. Thus, the informational value of rejection would also be fairly high – say, $p_R(\omega_3) = 0.4$. Finally, consider proposition ω_4, that you see a real yeti. Retaining that proposition would be of great value, for then your photo might make the newspapers, and you could receive scientific acclaim. Thus, the informational value of rejecting this proposition would be quite low – say, $p_R(\omega_4) = 0.07$.

To complete the description, we need to specify the probabilities associated with the four propositions. Supposing you are an experienced hiker who has never been the victim of altitude sickness, you might set the probability of hallucinating to be quite low – say $p_S(\omega_1) = 0.10$. However, it is reasonable to suppose that the probability that what you see is just another human being, perhaps rather large and scruffy, would be quite high, say $p_S(\omega_2) = 0.8$. Also, the chances that a practical joker is trying to trick you would be rather low, say $p_S(\omega_3) = 0.09$. Finally, since conclusive evidence that yeti exist has never been supplied and you would be perhaps the first to do so, you consider that to be highly unlikely, thus you would set the probability of what you observe being Yeti to be very low, say $p_S(\omega_4) = 0.01$.

An optimizing truth-seeker must choose the single proposition with the largest expected epistemic utility, which is ω_2. The corresponding probability of error, using (6.2), is $P_{error} = 0.2$.

An optimizing error-avoider, on the other hand, would compute the largest set to be rejected. Setting $q = 1$, the seriously possible set is $\Sigma_1 = \{\omega_1, \omega_2\}$, with the probability of error, using (6.3), easily seen to be $P_{error} = 0.13$. The price for reducing the probability of error is to reject fewer propositions, resulting in less informational value being retained.   □

Of course, to make a decision, one must ultimately settle on a unique element in Ω, and in that sense the above procedure is not a complete decision. Rather, it is a mechanism to eliminate bad choices, leaving only good choices. In that sense, any of the nonrejected propositions possesses a claim to being worthy of being chosen. Thus, although this approach may not result in a unique choice, it at least provides a way to narrow down the choices for further investigation.

Such further investigation can take different forms. One conservative approach is to suspend further judgment between the elements of Σ_q and seek further information in order to refine the set. Another approach is to renormalize the the probabilities by defining

$$P_S^1(B) = \frac{P_S(B)}{P_S(\Sigma_q)}$$

$$P_R^1(B) = \frac{P_R(B)}{P_R(\Sigma_q)}$$

for $B \in \Sigma_q$. We may then compute the iterated set

$$\Sigma_q^1 = \{\omega \in \Sigma_q : p_S^1(\omega) \geq q p_R^1(\omega)\},$$

resulting in a refined set of serious possibilities.

Another way to make a final choice is to select from Σ_q according to some ancillary criteria. One possibility is to choose the element of Σ_a that has the highest credal probability. Another possibility is to choose the element that has the lowest informational value of rejection. Yet a third possibility is to choose the one element for which the difference between the credal probability and q times the informational value of rejection is the greatest. Fortunately, all of these elements are available by inspection of Σ_q; thus, additional computation is minimized.

Still another way to refine the set Σ_q is to retain only those elements that are strictly seriously possible, which leads to the following definition:

Definition 6.1 Let $\Sigma_q = \{\omega \in \Sigma_q : p_S(\omega) \geq q p_R(\omega)\}$. $\omega \in \Sigma_q$ is *strictly seriously possible* if there exists no $\omega' \in \Sigma_q$ such that

$$p_R(\omega') < p_R(\omega) \text{ and } p_S(\omega') \geq p_S(\omega)$$

or

$$p_R(\omega') \leq p_R(\omega). \text{ and } p_S(\omega') > p_S(\omega).$$

□

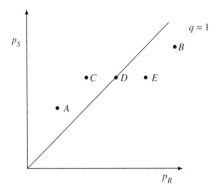

Figure 6.1. Cross-plot of credal and information-determining mass functions.

That is, ω is strictly seriously possible if there are no propositions with lower informational value of rejection and greater than or equal credibility, or that are more credible with less than or equal informational value of rejection. To illustrate, consider Figure 6.1. This figure is a cross-plot of p_R as the abscissa and p_S as the ordinate for a five-element set of propositions $\Omega = \{A, B, C, D, E, F\}$. The diagonal line represents the boundary of Σ_q for $q = 1$. By inspection, we see that the set of seriously possible propositions is $\{A, C, D\}$, but only A and C are strictly seriously possible.

Finally, a last way to refine the set is to increase q. Examination of Figure 6.1 indicates that by increasing q so that the line passes through A, all but that option would be eliminated. Thus, there are various ways by which the original set Σ_q can be refined in order to identify a unique choice. The advantage of computing the entire set, however, is that that the decision maker is able to formulate a comprehensive picture involving multiple criteria. This capability is similar in spirit to the Pareto frontier that arises in multiobjective decision theory. Borrowing from the terminology of that discipline, we may refer to the set of strictly seriously possible propositions as the *serious possibility frontier*.

It is important to emphasize that the inductive and abductive aspects of the problem must not be conflated. That is, the determination of credal probability must not be influenced by informational value. If it were, then we might be guilty of letting the desirability of a proposition influence our assessment of its veracity. Such behavior would be tantamount to wishful thinking.

6.1.2 Extension to the multivariate case

Epistemic utility theory was originally developed by Levi (1980) for a single agent. However, it easily extends to the multiple agent case. Let Ω_1 and Ω_2 be two sets of propositions with associated Boolean algebras $\mathcal{F}_1$ and $\mathcal{F}_2$, respectively. A *measurable rectangle* is a set of the form $B_1 \times B_2$, where $B_1 \in \mathcal{F}_1$ and $B_2 \in \mathcal{F}_2$. Elements of $B_1 \times B_2$ are the ordered pairs (ω_1, ω_2), where $\omega_1 \in B_1$ and $\omega_2 \in B_2$. Let $\mathcal{F}_1 \otimes \mathcal{F}_2$ denote the smallest Boolean algebra over the product space $\Omega_1 \times \Omega_2$ that contains all measurable rectangles.

Let P_{S_1} and P_{S_2} be expectation-determining probabilities defined over $\mathcal{F}_1$ and $\mathcal{F}_2$, respectively, and let P_{R_1} and P_{R_2} be the associated information-determining probabilities. Now let the *joint expectation-determining probability measure* $P_{S_{12}}$ and the *joint information-determining probability measure* $P_{R_{12}}$ be defined over $\mathcal{F}_1 \otimes \mathcal{F}_2$ such that

$$P_{S_1}(B_1) = P_{S_{12}}(B_1 \times \Omega_2)$$

$$P_{S_2}(B_2) = P_{S_{12}}(\Omega_1 \times B_2)$$

$$P_{R_1}(B_1) = P_{R_{12}}(B_1 \times \Omega_2)$$

$$P_{R_2}(B_2) = P_{R_{12}}(\Omega_1 \times B_2).$$

We thus have

$$p_{S_1}(\omega_1) = \sum_{\omega_2} p_{S_{12}}(\omega_1, \omega_2)$$

$$p_{S_2}(\omega_2) = \sum_{\omega_1} p_{S_{12}}(\omega_1, \omega_2)$$

$$p_{R_1}(\omega_1) = \sum_{\omega_2} p_{R_{12}}(\omega_1, \omega_2)$$

$$p_{R_2}(\omega_2) = \sum_{\omega_1} p_{R_{12}}(\omega_1, \omega_2),$$

where $p_{S_{12}}$ and $p_{R_{12}}$ are the joint mass functions associated with $P_{S_{12}}$ and $P_{R_{12}}$, respectively.

We may define the expected epistemic utility of any element $B_{12} \in \mathcal{F}_1 \otimes \mathcal{F}_2$ as

$$E\phi^q(B_{12}) = P_{S_{12}}(B_{12}) - q_{12}P_{R_{12}}(B_{12})$$

$$= \sum_{(\omega_1, \omega_2) \in B_{12}} p_{S_{12}}(\omega_1, \omega_2) - q_{12}p_{R_{12}}(\omega_1, \omega_2),$$

and the set of jointly nonrejected propositions is

$$\Sigma_{q_{12}} = \{(\omega_1, \omega_2) \in \Omega_1 \times \Omega_2 \colon p_{S_{12}}(\omega_1, \omega_2) \geq q_{12}p_{R_{12}}(\omega_1, \omega_2)\},$$

and the sets of marginally nonrejected propositions are

$$\Sigma_{q_1} = \{\omega_1 \in \Omega_1 \colon p_{S_1}(\omega_1) \geq q_1 p_{R_1}(\omega_1)\}$$
$$\Sigma_{q_2} = \{\omega_2 \in \Omega_2 \colon p_{S_2}(\omega_2) \geq q_2 p_{R_2}(\omega_2)\}.$$

We say that a joint decision problem is *informationally independent* if $P_{R_{12}}(B_1 \times B_2) = P_{R_1}(B_1)P_{R_2}(B_2)$ for all measurable rectangles $B_1 \times B_2$ and *credally independent* if $P_{S_{12}}(B_1 \times B_2) = P_{S_1}(B_1)P_{S_2}(B_2)$ for all measurable rectangles $B_1 \times B_2$. If the joint problem is both informationally and credally independent, then, provided $q_1 q_2 \geq q_{12}$,

$$\Sigma_{q_1} \times \Sigma_{q_2} \subset \Sigma_{q_{12}}.$$

To establish this result, we simply note that if $\omega_1 \in \Sigma_{q_1}$ and $\omega_2 \in \Sigma_{q_2}$, then

$$p_{S_{12}}(\omega_1, \omega_2) = p_{S_1}(\omega_1)p_{S_2}(\omega_2)$$
$$\geq q_1 q_2 p_{R_1}(\omega_1)p_{R_2}(\omega_2)$$
$$\geq q_{12} p_{R_{12}}(\omega_1, \omega_2).$$

Thus, $(\omega_1, \omega_2) \in \Sigma_{q_{12}}$.

In the absence of independence, the relationship between $\Sigma_{q_1} \times \Sigma_{q_2}$ and $\Sigma_{q_{12}}$ is more difficult to determine. However, we can establish the following theorem.

Theorem 6.1 *Let Ω_1 and Ω_2 be disjoint proposition sets Let $q_{12} \leq q_1$. Then, if $\omega_1 \in \Sigma_{q_1}$, there exists an ω_2^* such that $(\omega_1, \omega_2^*) \in \Sigma_{q_{12}}$.*

Proof: This result is established by the contrapositive, namely, that if $(\omega_1, \omega_2) \notin \Sigma_{12}$ for all $\omega_2 \in \Omega_2$, then $\omega_1 \notin \Sigma_{q_1}$. Suppose

$$p_{S_{12}}(\omega_1, \omega_2) < q_{12} p_{R_{12}}(\omega_1, \omega_2) \quad \text{for all } \omega_2 \in \Omega_2.$$

Then

$$p_{S_1}(\omega_1) = \sum_{\omega_2} p_{S_{12}}(\omega_1, \omega_2)$$
$$< q_{12} \sum_{\omega_2} p_{R_{12}}(\omega_1, \omega_2)$$
$$= q_{12} p_{R_1}(\omega_1)$$
$$\leq q_1 p_{R_1}(\omega_1);$$

hence, $\omega_1 \notin \Sigma_{q_1}$. $\qquad\qquad\square$

The content of this theorem is that if ω_1 is a serious possibility in Ω_1, then there exists an $\omega_2^* \in \Omega_2$ such that the joint proposition $(\omega_1, \omega_2^*) \in \Sigma_{q_1} \times \Omega_2$ lies in $\Sigma_{q_{12}}$.

6.2 Failure avoidance

One who seeks the best and only the best solution is similar to one who seeks to know the truth and nothing but the truth. The former is a optimizer, and the latter is an idealist. Both, however, must sometimes compromise in the face of conflicting or insufficient information.

The classical epistemological definition of knowledge is true justified belief, and the focus is on justification, that is, identifying the truth and nothing but the truth. In contrast, Levi (1980) argues that one's current knowledge corpus is what it is, and rather than concentrating on justification, one should concentrate on devising ways to improve it. To pursue this objective, Levi proposes, as developed in Section 6.1, a more conservative approach: *acquiring information while avoiding error*, as implemented via his epistemic utility theory.

Epistemology involves the consideration of propositions in terms of knowledge and belief. A parallel concept to epistemology is praxeology, which involves the consideration of propositions in terms of their expedience and efficiency. Thus, while the classical praxeological decision-making focus is on taking the best and only the best action, Levi (1980) invites consideration of an alternative praxeological approach: *Conserving resources while avoiding failure*.

Let us develop this analogue further. In the epistemological context, informational value is a resource. It is consumed by retaining propositions, and it is conserved by rejecting them. Thus, if a proposition has high informational value of rejection, that is, if $p_R(\omega) \approx 1$, then by rejecting ω we bank $p_R(\omega)$ worth of informational value – it will never be consumed. If, on the other hand, $p_R(\omega) \approx 0$, then we would consume very little informational value by retaining it. Also, the probability of error is expressed in terms of p_S: the smaller $p_S(\omega)$, the more likely ω is to be false. Thus, the test given by (6.1) retains only those propositions whose chances of error are small relative to the informational value they provide.

In the praxeological context, the resources available to the decision maker can take several forms. For example, they can be economic, political, moral, aesthetic, or personal. They may involve monetary costs, damage to material, exposure to hazard, social reputation, consumption of energy, risk to personnel, and so forth. Let us interpret the utility p_R as an inefficiency measure: The larger $p_R(\omega)$, the more costly the consequence. Also, let us interpret the utility p_S as an expedience measure: The larger $p_S(\omega)$, the more ω avoids failure. Thus, an analogue to the epistemological test would be to retain only those consequences whose propensity to failure are small relative to the resources they conserve. Failure avoidance, however, often admits a relative rather than an absolute interpretation, and is expressible by degree. A key point is that, while there may be only one choice that maximizes a performance metric, there may be many choices that sufficiently avoid failure and provide adequate performance. In this regard, the focus thus turns to how to establish a meaningful notion of adequacy.

6.2.1 Multiple selves

To consider and think over rationally the advantages or benefits I would gain by holding …the proposed position or benefice, and on the other hand to consider in the same way the disadvantages and dangers in holding it. Do the same with the alternative; look at the advantages and benefits of not holding it, and conversely the disadvantages and dangers in not holding it.

After having thought over and reflected in this way from every point of view on the matter before me, I shall look to which side reason most inclines, and thus it is according to the stronger movement of the reason, and not through any sensual inclination, that one should make up one's mind on the matter before one.

— *Ignatius of Loyola*
The Spiritual Exercises, 181, 182 (1548)

Many theorists (see Elster, 1985; Harsanyi, 1955; Margolis, 1990) have argued that it is unwise to aggregate conflicting interests into a single preference ordering. Some have asserted that in a social setting, individuals have multiple selves. These selves are similar to the "facets" or "aspects" of an individual, as defined by Steedman and Krause (1985), who maintain that an individual, although an indivisible unit, is nevertheless capable of considering its choices from different points of view, and that separate utilities may be defined to correspond to each facet of an individual.

There are many points of view that might be considered: egoism versus altruism, group versus individual, risk proneness versus risk averseness, and so forth. Such comparisons, however, are largely manifestations of the same underlying attribute. Consider, for example, egoism versus altruism. Both attributes are concerned with whose interests are primary. Munificence and avarice are extremes on the same sociological attribute scale. Also, how one views what is good for the individual in relation to what is good for the group can be placed somewhere on a scale between two extremes. The same can be said for one's position on risk and for many other ways to define different points of view.

To be truly different facets, however, the points of view ought to be fundamentally unrelated in that one cannot use one point of view to define the other one. Two such facets for which this is not possible are effectiveness and efficiency. Effectiveness is a measure of how successfully a choice achieves some objective without regard for how much it costs, and and efficiency is a measure of how much resource (e.g., cost, time) is consumed by making the choice without concern for how well the choice achieves the objective. Of course, it may turn out that an effective choice is also a costly one, but that relationship is not implied. It is also possible for an effective choice to be inexpensive, for an ineffective choice to be expensive, and for an ineffective choice to be inexpensive.

Effectiveness and efficiency are fundamental praxeological concepts that permit a fairly comprehensive evaluation of behavior. Although it is possible to consider other notions, such as the legality, ethicality, ecological soundness, and social responsibleness of a choice, it is often the case that these attributes can be subsumed into one of the fundamental praxeological facets. We do not claim that all decision scenarios can be couched in terms of effectiveness and efficiency, but we do assert that a great many interesting problems can be reduced to evaluations of two main issues: "How good of a job will it do?" and "How much does it cost?" The point of this discussion is that, rather than attempting to combine these two aspects, let us see what emerges if we keep them distinct.

We shall view each stakeholder X_i as being composed of two selves: the *selecting self*, denoted S_i, which evaluates consequences in terms of effectiveness (avoiding failure) without concern for efficiency, and the *rejecting self*, denoted R_i, which evaluates consequences in terms of efficiency (conserving resources), without concern for effectiveness. Essentially, these selves are the atoms of the group. Decomposing the society into selves provides a natural way to account for the inner conflict that individuals may experience. When viewed simultaneously from both perspectives, the individual is represented as the concatenation of these two selves, that is, $X_i \equiv S_i R_i$, which simultaneously takes into consideration both effectiveness and efficiency.

Partitioning the individual into the selves permits a more comprehensive and informed consideration of its preferences. An optimizer can only optimize one thing at a time, and so must somehow combine the effectiveness and efficiency measures into a single utility. This requires the stakeholder to form an implicit compromise between its two selves. Thus, in a very real sense, even a single individual can be viewed as a multicriterion decision maker whose individual self utilities must be aggregated.

6.2.2 Single-stakeholder satisficing

Simon (1955) has appropriated the term *satisficing* to mean "good enough" and advocates the construction of "aspiration levels" of how good a solution might reasonably be achieved, and halting the search when these expectations are met. Although aspiration levels at least superficially establish minimum requirements, this approach relies primarily on experience-derived expectations. If the aspiration is too low, performance may needlessly be sacrificed, and if it is too high, there may be no solution. It is difficult to establish a good and practically attainable aspiration level without first exploring the limits of what is possible, or, in other words, without first identifying optimal solutions – the very activity that satisficing is intended to circumvent. Furthermore, this concept of being good enough actually begs the question, because being good enough requires us to define minimum requirements (aspirations), which immediately plunges us into a tautology, since the criterion for defining minimum requirements can only be that they are good enough.

The failure avoidance framing motivates a new definition of the term *satisficing*. We retain the "satisficing" terminology[4] because our usage is consistent with the issue that motivated Simon's (1955) original usage – to identify options that are good enough in the sense of comparing attributes of the options to a standard. Our usage differs only in the standard used for comparison. Whereas Simon's approach involves heuristic considerations to define a rational solution,[5] our failure avoidance framing motivates the development of a mathematically precise definition of satisficing rationality.

Definition 6.2 A choice is *satisficingly rational* if the gains achieved by making the choice are equal to or exceed the losses, provided the gains and losses can be expressed in commensurable units. □

The degree to which an action results in a gain (i.e., avoids failure) is termed its *selectability* and the degree to which it creates a loss (i.e., consumes resources) is termed its *rejectability* We adopt the more modest concept of avoiding failure, rather than success, in order to emphasize the different point of view associated with satisficing decision theory.

---

[4] Other researchers have appropriated this term to describe various notions of constrained optimization. In this work, however, usage is restricted to be consistent with Simon's (1955) original concept.

[5] Simon has coined the term "bounded rationality" to account for the fact that, in practical situations, the lack of sufficient information and computational power prohibit a decision maker from achieving an optimal solution; hence, heuristic procedures must be invoked to define a decision that is good enough.

We now consider how to define a satisficing decision for a single stakeholder. Let $\mathcal{A}$ be the action space for stakeholder X with selecting self S and rejecting self R.

Definition 6.3 The *selectability utility* u_S is a mass function over $\mathcal{A}$ that characterizes the effectiveness of elements $\mathcal{A}$ with respect to X avoiding failure. The *rejectability utility* u_R is a mass function over $\mathcal{A}$ that characterizes the efficiency of elements $\mathcal{A}$ with respect to consuming X's resources.   □

The relationship $u_S(a) > u_S(a')$ means that a is more effective than a' in terms of avoiding failure. Similarly, the relationship $u_R(a) > u_R(a')$ means that a is less efficient than a' in terms of conserving resources.

Definition 6.4 Let u_S and u_R be selectability and rejectability utilities for X, and let $0 < q \leq 1$. The set

$$\Sigma_q = \{a \in \mathcal{A}: u_S(a) \geq q u_R(a)\} \qquad (6.4)$$

is called a *satisficing set* for X at boldness level q.   □

It is straightforward to see that if $q \leq 1$, then $\Sigma_q \neq \varnothing$. To establish this claim, suppose $u_S(a) < q u_R(a)$ for all $a \in \mathcal{A}$. Then

$$1 = \sum_a u_S(a) < q \sum_a u_R(a) = q \,,$$

a contradiction. Thus, Σ_q must contain at least one element.

The satisficing set constitutes the set of actions for which effectiveness, as measured by the selectability utility, is at least as great as q times the inefficiency, as measured by the rejectability utility. The praxeological interpretation of q is a measure of boldness, as with the epistemic context. As q increases, so does the number of actions that are rejected, indicating that the stakeholder is increasingly willing to risk failure in the interest of conserving resources.

The classical optimization-based framing and the failure-avoidance framing of a decision problem differ fundamentally. Whereas the classical framing involves comparisons of a single (composite) attribute between *different actions* to identify the best one, the failure avoidance framing involves comparisons between *different attributes for each action* to decide whether or not to reject the action. Colloquially, the classical framing can be interpreted as insisting on "the best and only the best," but the failure-avoidance framing can be interpreted as being content with "getting what you paid for." A satisficing decision maker is one who seeks a compromise between effectiveness and efficiency.

Although the satisficing formulation is more modest in its goal than the classical optimizing formulation based on the best-and-only-the-best paradigm, satisficing decisions can be viewed as optimal: They eliminate the maximum possible number of potential consequences, subject to the constraint of avoiding failure. *Satisficing decision makers are optimal failure-avoiders.* Furthermore, if they succeed in eliminating all but one action, they will become optimizers in the classical sense. (As Stirling (2003) has shown, an optimal solution is also a satisficing solution.) Thus, rather than a heuristic approximation to classical optimization, *satisficing as presented here is a generalization of classical optimization.* If warranted by the evidence, only a truly optimal solution will survive the process of elimination upon which the satisficing concept rests.

To ensure a well-formed decision problem, the concepts of effectiveness and efficiency must not be restatements of the same attribute. Consequently, for a single-stakeholder decision problem, it is reasonable to assume that the selectability of an attribute should not depend on its rejectability, or vice versa. In other words, a fundamental assumption in a single-stakeholder context is that the selecting self and the rejecting self are praxeologically independent; that is, if we consider the concordant utility U_{SR}, it must hold that

$$U_{S|R}(a,a') = u_S(a)u_R(a') \ \forall \ a,a' \in \mathcal{A}. \tag{6.5}$$

In terms of conditional utilities, this means that $u_{S|R}(a|a') = u_S(a) \ \forall \ a,a' \in \mathcal{A}$ and $u_{R|S}(a|a') = u_R(a) \ \forall \ a,a' \in \mathcal{A}$.

6.2.3 Multiple stakeholder satisficing

For single-agent, low-dimensional problems, Simon's heuristic approach of specifying an aspiration level may be noncontroversial. But with multiagent systems, interdependence between decision makers can be complex, and aspiration levels can be conditional (what is good enough for me may depend on what is good enough for you). The current state of affairs regarding aspiration levels does not appear to address completely the problem of specifying minimum requirements in multistakeholder contexts.

The real power of our satisficing approach becomes evident when we extend to the multiple-agent case, in which, as we have previously noted, it is not generally possible for a group and all of its members to achieve simultaneously optimal solutions. However, as we shall establish, it is possible for a group and all its members to obtain optimal failure-avoiding solutions. Furthermore, if a classically optimal solution exists, it will also be a satisficing solution.

We may view an n-member group $\mathcal{X}_n = \{X_1, \ldots, X_n\}$ of individual stakeholders as a $2n$-member group comprising an n-member subgroup of selecting selves $\mathcal{S}_n = \{S_1, \ldots, S_n\}$ and an n-member subgroup of rejecting selves $\mathcal{R}_n = \{R_1, \ldots, R_n\}$. Expressed in terms of the selves, the group can be expressed as

$$\mathcal{X}_n \equiv \mathcal{S}_n \mathcal{R}_n = \{S_1, \ldots, S_n, R_1, \ldots, R_n\}.$$

To develop a theory of multistakeholder decision making in the satisficing context, we invoke the principles established in Chapter 2 and assume that they apply to the atoms (the selecting and rejecting selves) rather than directly to the stakeholders. We may construct a utility network whose nodes are the selecting and rejecting selves, and whose edges are conditional utilities characterizing the social influence relationships.

Two conjecture profiles are associated with each stakeholder – one for the selecting self and one for the rejecting self. Let $\mathbf{s}_i \in \mathcal{A}$ denote the conjecture of the selecting self S_i, and let $\mathbf{r}_i \in \mathcal{A}$ denote the conjecture of the rejecting self R_i. Now suppose S_i has p_i selecting parents and z_i rejecting parents, that is,

$$\mathrm{pa}\,(S_i) = \{S_{i_1}, \ldots, S_{i_{p_i}}, R_{k_1}, \ldots, R_{k_{z_i}}\}.$$

Also, suppose R_j has t_j selecting parents and v_j rejecting parents, that is,

$$\mathrm{pa}\,(R_j) = \{S_{j_1}, \ldots, S_{j_{t_j}}, R_{l_1}, \ldots, R_{l_{v_j}}\}.$$

Let $\{\mathbf{s}_{i_1}, \ldots, \mathbf{s}_{i_{p_i}}, \mathbf{r}_{k_1}, \ldots, \mathbf{r}_{k_{z_i}}\}$ and $\{\mathbf{s}_{j_1}, \ldots, \mathbf{s}_{j_{t_j}}, \mathbf{r}_{l_1}, \ldots, \mathbf{r}_{l_{v_j}}\}$ denote the joint conjectures of $\mathrm{pa}\,(S_i)$ and $\mathrm{pa}\,(R_j)$, respectively. The concordant utility thus becomes

$$U_{S_1 \cdots S_n R_1 \cdots R_n}(\mathbf{s}_1, \ldots, \mathbf{s}_n, \mathbf{r}_1, \ldots, \mathbf{r}_n) = \prod_{i=1}^{n} u_{S_i | \mathrm{pa}(S_i)}(\mathbf{s}_i | \mathbf{s}_{k_1}, \ldots, \mathbf{s}_{k_{p_i}}, \mathbf{r}_{m_1}, \ldots, \mathbf{r}_{m_{z_i}})$$

$$\prod_{j=1}^{n} u_{R_j | \mathrm{pa}(R_j)}(\mathbf{r}_j | \mathbf{s}_{l_1}, \ldots, \mathbf{s}_{l_{t_j}}, \mathbf{r}_{l_1}, \ldots, \mathbf{r}_{l_{v_j}}).$$

(6.6)

Consider the network of the three-agent (and thus six-atom) group displayed in Figure 6.2. By inspection we see that $\mathrm{pa}\,(S_2) = \mathrm{pa}\,(S_3) = \mathrm{pa}\,(R_2) = \{S_1\}$, $\mathrm{pa}\,(R_1) = \{S_2\}$, and $\mathrm{pa}\,(R_3) = \{S_2, R_2\}$; thus, the concordant utility is

$$U_{S_1 S_2 S_3 R_1 R_2 R_3}(\mathbf{s}_1, \mathbf{s}_2, \mathbf{s}, \mathbf{r}_1, \mathbf{r}_2, \mathbf{r}_3) = u_{S_1}(\mathbf{s}_1) u_{S_2 | S_1}(\mathbf{s}_2 | \mathbf{s}_1) u_{S_3 | S_1}(\mathbf{s}_3 | \mathbf{s}_1)$$

$$u_{R_1 | S_2}(\mathbf{r}_1 | \mathbf{s}_2) u_{R_2 | S_1}(\mathbf{r}_2 | \mathbf{s}_1) u_{R_3 | S_2 R_2}(\mathbf{r}_3 | \mathbf{s}_2, \mathbf{r}_2).$$

Recall that the acyclicity principle stipulates that there can be no paths in the utility network such that a stakeholder can influence itself. We also stipulate, via (6.5), that in the single-stakeholder case, the selecting agent cannot influence the rejecting agent, and vice versa. However, in the multistakeholder case, it is possible that a stakeholder can influence itself through paths that involve other agents. We see in Figure 6.2 that although S_1 influences R_1, it does so through S_2 without creating a cycle. Thus, it is possible that one stakeholder self can

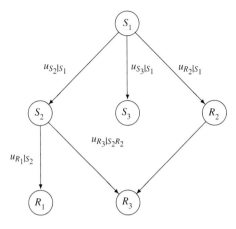

Figure 6.2. A satisficing network for a three-stakeholder group.

indirectly influence its other self by influencing other stakeholders who in turn influence the other self.

This network displayed in Figure 6.2 also illustrates how the acyclicity constraint can be partially overcome. Recall that acyclicity means that, if X_1 directly influences X_2, then X_2 may not directly influence X_1. By representing a society in terms of selecting and rejecting selves, however, this constraint can be relaxed, since, as the figure illustrates, S_1 directly influences R_2 and, simultaneously, S_2 directly influences R_1 without violating the acyclicity constraint.

As mentioned in Chapter 2, however, the concordant utility does not directly serve as the basis for taking action, because it is a function of multiple profiles (conjectures), and only one profile can actually be implemented. Rather, its function is to encode all of the interrelationships that exist among the atoms. To illustrate, let us consider the single-stakeholder X, and therefore the two-self problem defined by $U_{SR}(a,a')$. This function represents the conflict that the individual experiences when viewing a from the perspective of effectiveness and at the same time viewing a' from the perspective of efficiency. Now suppose $a = a'$. By the assumption of praxeic independence for an individual, $U_{SR}(a,a) = u_S(a)u_R(a)$. If both $u_S(a)$ and $u_R(a)$ are large, then a is both effective and inefficient. Thus, the individual is highly conflicted, because the action is desirable from the point of view of achieving the goal but undesirable because it is a large consumer of resources. If either factor is small, then the product is small and there is little conflict: Either the action is not effective, so regardless of cost it is not desirable from that point of view, or it is cheap, and so it is desirable from that point of view, regardless of how well it achieves the goal.

6.3 Satisficing games

To attain knowledge, add things every day,
To attain wisdom, subtract things every day.

— *Lao Tzu*
Tao Tê Ching, XLVIII

By introducing the concept of two selves as a refinement of an individual – one concerned with the effectiveness and the other with the efficiency of each feasible action – we open the possibility of developing a game-theoretic approach to decision making that takes advantage of this refinement. We thus introduce the notion of a conditional satisficing game.

Definition 6.5 Let $\mathcal{X}_n = \{X_1, \ldots, X_n\}$ be an n-member group with joint action space $\mathcal{A} = \mathcal{A}_1 \times \cdots \times \mathcal{A}_n$, selecting selves $\mathcal{S}_n = \{S_1, \ldots, S_n\}$, and rejecting selves $\mathcal{R}_n = \{R_1, \ldots, R_n\}$. A *conditional satisficing game* is a quadruple $\{\mathcal{S}_n, \mathcal{R}_n, \mathcal{A}, U_{\mathcal{S}_n \mathcal{R}_n}\}$, where $U_{\mathcal{S}_n \mathcal{R}_n}$ is the concordant utility given by (6.6). □

Even though the game is expressed in terms of the selecting and rejecting selves, S_i and R_i, respectively, these selves are not empowered to act; they simply represent facets of X_i. Thus, to solve a satisficing game we must define a solution concept for X_i. To do so, we proceed similarly to the way we defined solutions for conditional games in Chapter 3.

Definition 6.6 Given a concordant utility $U_{\mathcal{S}_n \mathcal{R}_n}$, let s_{ij} and r_{ij} denote the jth elements of $\mathbf{s}_i$ and $\mathbf{r}_i$, respectively, form the profiles $(s_{11}, \ldots, s_{nn})$ and $(r_{11}, \ldots, r_{nn})$, and sum the concordant utility over all elements of each $\mathbf{s}_i$ and $\mathbf{r}_i$ except the iith elements to form the *satisficing group welfare function*

$$v_{\mathcal{S}_n \mathcal{R}_n}(s_{11}, \ldots, s_{nn}, r_{11}, \ldots, r_{nn}) = \sum_{\sim s_{11}} \cdots \sum_{\sim s_{nn}} \cdots \sum_{\sim r_{11}} \cdots$$
$$\sum_{\sim r_{nn}} U_{\mathcal{S}_n \mathcal{R}_n}(\mathbf{s}_1, \ldots, \mathbf{s}_n, \mathbf{r}_1, \ldots, \mathbf{r}_n). \quad (6.7)$$

□

We observe that if the group is completely dissociated (see Definition 3.9), then the satisficing group welfare function coincides with the concordant utility.

Definition 6.7 Given a satisficing group welfare function $v_{\mathcal{S}_n \mathcal{R}_n}(s_1, \ldots, s_n, r_1, \ldots, r_n)$, the *joint selectability function* is obtained by summing the satisficing group welfare function over $r_1, \ldots, r_n$, yielding

$$v_{\mathcal{S}_n}(s_1, \ldots, s_n) = \sum_{r_1, \ldots, r_n} v_{\mathcal{S}_n \mathcal{R}_n}(s_1, \ldots, s_n, r_1, \ldots, r_n). \qquad (6.8)$$

Similarly, the *joint rejectability function* is obtained by summing the satisficing group welfare function over $s_1, \ldots, s_n$, yielding

$$v_{\mathcal{R}_n}(r_1, \ldots, r_n) = \sum_{s_1, \ldots, s_n} v_{\mathcal{S}_n \mathcal{R}_n}(s_1, \ldots, s_n, r_1, \ldots, r_n). \qquad (6.9)$$

□

Given the joint selectability and rejectability functions, we may compute the set of action profiles that are *jointly satisficing* for the group $\mathcal{X}_n$.

Definition 6.8 The *jointly satisficing set* is

$$\Sigma_{\mathcal{X}_n} = \{\mathbf{a} \in \mathcal{A} \colon v_{\mathcal{S}_n}(\mathbf{a}) \geq q_{\mathcal{X}_n} v_{\mathcal{R}_n}(\mathbf{a})\},$$

where $q_{\mathcal{X}_n}$ is the q-value for the group. □

The jointly satisficing set is the set of action profiles that are good enough in the sense that the degree of concordance among the members of the group for selecting any element is at least as great as the boldness factor times the cost of resources.

We may also compute the selectability and rejectability functions for subgroups.

Definition 6.9 Let $\mathcal{X}_k = \{X_{i_1}, \ldots, X_{i_k}\}$ be a k-element subgroup of $\mathcal{X}_n$ and let $\mathcal{S}_k = \{S_{i_1}, \ldots, S_{i_k}\}$ and $\mathcal{R}_k = \{R_{i_1}, \ldots, R_{i_k}\}$ denote the selecting and rejecting subgroups, respectively. The *subgroup selectability and rejectability functions* are the corresponding k-dimensional marginals of the selectability and rejectability functions, yielding

$$v_{\mathcal{S}_k}(a_{i_1}, \ldots, a_{i_k}) = \sum_{\sim \{a_{i_1}, \ldots, a_{i_k}\}} v_{\mathcal{S}_n}(a_1, \ldots, a_n)$$

and

$$v_{\mathcal{R}_k}(a_{i_1}, \ldots, a_{i_k}) = \sum_{\sim \{a_{i_1}, \ldots, a_{i_k}\}} v_{\mathcal{R}_n}(a_1, \ldots, a_n).$$

The *subgroup satisficing set* is

$$\Sigma_{\mathcal{X}_k} = \left\{ (a_{i_1}, \ldots, a_{i_k}) \in \mathcal{A}_{i_1} \times \cdots \times \mathcal{A}_{i_k} \colon v_{\mathcal{S}_k}(a_{i_1}, \ldots, a_{i_k}) \geq q_{\mathcal{X}_k} v_{\mathcal{R}_k}(a_{i_1}, \ldots, a_{i_k}) \right\}.$$

□

When $k = 1$, the *individual selectability and rejectability functions* are the marginals of $v_{\mathcal{S}_n}$ and $v_{\mathcal{R}_n}$, yielding

$$v_{S_i}(a_i) = \sum_{\sim a_i} v_{\mathcal{S}_n}(a_1, \ldots, a_n)$$

and

$$v_{R_i}(a_i) = \sum_{\sim a_i} v_{\mathcal{R}_n}(a_1, \ldots, a_n)$$

for $i = 1, \ldots, n$. The *individually satisficing sets* are

$$\Sigma^i_{q_{X_i}} = \{a_i \in \mathcal{A}_i : v_{S_i}(a_i) \geq q_{X_i} v_{R_i}(a_i)\}$$

for $i = 1, \ldots, n$. Notice that each stakeholder may have its own q-value.

Definition 6.10 Let $\Sigma_{q_{X_i}}$, $i = 1, \ldots, n$ be the individually satisficing sets for a group $\mathcal{X}_n$. The *satisficing rectangle*

$$\mathcal{R}_{\mathcal{X}_n} = \Sigma^1_{q_{X_1}} \times \cdots \times \Sigma^n_{q_{X_n}}$$

is the Cartesian product of the individually satisficing sets. □

The satisficing rectangle is the set of all action profiles that are simultaneously satisficing for all of the stakeholders. It differs from the jointly satisficing set in that the satisficing rectangle is defined in terms of the individually satisficing sets, which are obtained from the marginals of the joint selectability and rejectability functions, whereas the jointly satisficing set is defined in terms of the joint selectability and rejectability functions. Consequently, the satisficing rectangle does not represent a group preference. In fact, the two sets can be disjoint.

6.4 Satisficing coherence

In Section 2.3 we established conditions for no stakeholder to be categorically subjugated, that is, a condition such that it cannot happen that all outcomes that are good for the group are bad for some individual. In this section, we develop an analogous result for the satisficing context.

Let $\mathcal{X}_n$ be a group, and let $\mathcal{X}_m = \{X_{j_1}, \ldots, X_{j_m}\}$ and $\mathcal{X}_k = \{X_{i_1}, \ldots, X_{i_k}\}$ be arbitrary disjoint subgroups with action subspaces $\mathcal{A}_m = \mathcal{A}_{j_1} \times \cdots \times \mathcal{A}_{j_m}$ and $\mathcal{A}_k = \mathcal{A}_{i_1} \times \cdots \times \mathcal{A}_{i_k}$, respectively. Now form the subgroup $\mathcal{X}_{mk}$ as the union of $\mathcal{X}_m$ and $\mathcal{X}_k$, that is,

$$\mathcal{X}_{mk} = \mathcal{X}_m \cup \mathcal{X}_k = \{X_{j_1}, \ldots, X_{j_m}, X_{i_1}, \ldots, X_{i_k}\}.$$

For $\mathbf{a} = (a_1, \ldots, a_n)$, define $\mathbf{a}_{mk} = (\mathbf{a}_m, \mathbf{a}_k)$, where $\mathbf{a}_m = (a_{j_1}, \ldots, a_{j_m})$ and $\mathbf{a}_k = (a_{i_1}, \ldots, a_{i_k})$. The jointly satisficing set for $\mathcal{X}_{mk}$ is

$$\Sigma_{\mathcal{X}_{mk}} = \{(\mathbf{a}_m, \mathbf{a}_k) \in \mathcal{A}_m \times \mathcal{A}_k : v_{\mathcal{S}_{mk}}(\mathbf{a}_m, \mathbf{a}_k) \geq q_{\mathcal{X}_{mk}} v_{\mathcal{R}_{mk}}(\mathbf{a}_m, \mathbf{a}_k)\}, \quad (6.10)$$

where

$$v_{\mathcal{S}_{mk}}(\mathbf{a}_m, \mathbf{a}_k) = \sum_{\sim(\mathbf{a}_m, \mathbf{a}_k)} v_{\mathcal{S}_n}(\mathbf{a}_n)$$

and

$$v_{\mathcal{R}_{mk}}(\mathbf{a}_m, \mathbf{a}_k) = \sum_{\sim(\mathbf{a}_m, \mathbf{a}_k)} v_{\mathcal{R}_n}(\mathbf{a}_n).$$

Now let us consider the relationship between the jointly satisficing set defined by (6.10) and the set formed as the Cartesian product $\Sigma_{\mathcal{X}_k \mathcal{X}_m} = \Sigma_{\mathcal{X}_k} \times \Sigma_{\mathcal{X}_m}$ of the jointly satisficing sets for the subgroups $\Sigma_{\mathcal{X}_m}$ and $\Sigma_{\mathcal{X}_k}$. The intersection $\Sigma_{\mathcal{X}_{mk}} \cap \Sigma_{\mathcal{X}_m \mathcal{X}_k}$ comprises the actions that are simultaneously jointly satisficing for the subgroup $\mathcal{X}_{mk}$ and also separately jointly satisficing for each of its components, $\mathcal{X}_m$ and $\mathcal{X}_k$. This set may be empty, but that is not a weakness of the theory. Rather, it is the recognition that societies can be populated by subgroups who are so diametrically opposed to each other that they cannot agree to do anything jointly satisficing. The following weak relationship does exist, however, between the two sets.

Theorem 6.2 *Let $\mathcal{X}_n$ be a group, and let $\mathcal{X}_m = \{X_{j_1}, \ldots, X_{j_m}\}$ and $\mathcal{X}_k = \{X_{i_1}, \ldots, X_{i_k}\}$ be arbitrary disjoint subgroups with action subspaces $\mathcal{A}_m = \mathcal{A}_{j_1} \times \cdots \times \mathcal{A}_{j_m}$ and $\mathcal{A}_k = \mathcal{A}_{i_1} \times \cdots \times \mathcal{A}_{i_k}$, respectively. If $\mathbf{a}_m \in \Sigma_{\mathcal{X}_m}$ and $q_{\mathcal{X}_{mk}} \leq q_{\mathcal{X}_m}$, then there exists $\mathbf{a}_k^* \in \mathcal{A}_k$ such that $(\mathbf{a}_m, \mathbf{a}_k^*) \in \Sigma_{X_{mk}}$.*

Proof: This result is established by the contrapositive, namely, that if $(\mathbf{a}_m, \mathbf{a}_k) \notin \Sigma_{X_{mk}}$ for all $\mathbf{a}_k \in \mathcal{A}_k$, then $\mathbf{a}_m \notin \Sigma_{\mathcal{X}_m}$. Suppose $\mathbf{a}_m \in \Sigma_{\mathcal{X}_m}$ but

$$v_{\mathcal{S}_{mk}}(\mathbf{a}_m, \mathbf{a}_k) < q_{\mathcal{X}_{mk}} v_{\mathcal{R}_{mk}}(\mathbf{a}_m, \mathbf{a}_k) \quad \text{for all } \mathbf{a}_k \in \mathcal{A}_k.$$

Then

$$v_{\mathcal{S}_m}(\mathbf{a}_m) = \sum_{\mathbf{a}_k} v_{\mathcal{S}_{mk}}(\mathbf{a}_m, \mathbf{a}_k)$$

$$< q_{\mathcal{X}_{mk}} \sum_{\mathbf{a}_k} v_{\mathcal{R}_{mk}}(\mathbf{a}_m, \mathbf{a}_k)$$

$$= q_{\mathcal{X}_{mk}} v_{\mathcal{R}_m}(\mathbf{a}_m)$$

$$\leq q_{\mathcal{X}_m} v_{\mathcal{R}_m}(\mathbf{a}_m);$$

hence, $\mathbf{a}_m \notin \Sigma_{\mathcal{X}_m}$, a contradiction. $\square$

Although this theorem is simple, it is important because it provides conditions such that the sure subjugation of $\mathcal{X}_m$ in order to accommodate the interests of the larger society is avoided. In other words, Theorem 6.2 establishes that every individual has a seat at the table in the sense that if an action is individually satisficing for it, then that action is an element of at least one jointly satisficing solution. This condition is perhaps the weakest one possible for meaningful negotiations to occur.

The ability of each decision maker to adjust its own q_{X_i} provides a negotiation mechanism by which the group can autonomously explore the effects of constraints on the decision problem. This is an important new capability. If a given set of constraints leads to a solution judged to be inadequate, conventional methodologies require the constraints to be revised by trial and error – their effects cannot be judged without generating a new solution. By explicitly representing the effects of social constraints on group decisions, satisficing game theory makes those constraints available for dynamic modification by the decision makers themselves, thus increasing the environmental variability with which the group can cope. For example, decision makers may resolve an impasse by relaxing their standards of performance. This may be done by each player incrementally reducing its q_{X_i} and recomputing the compromise set until it becomes nonempty. If the set remains empty after each decision maker has reduced its q_{X_i} to its minimum acceptable level, then an impasse cannot be avoided. Such a society is dysfunctional.

6.5 Satisficing compromises

The set $\mathbf{C}_{\mathcal{X}_n} = \mathcal{R}_{\mathcal{X}_n} \cap \Sigma_{\mathcal{X}_n}$ corresponds to the subset of action profiles that balance the interests of the individuals with the interests of the group and comprises the set of all action profiles that are simultaneously acceptable to the group and to each member. Assuming (perhaps as a result of negotiation) that it is nonempty, there is no guarantee that it is a singleton; there may be multiple compromise decisions. This set can be refined by first eliminating any satisficing solutions that are dominated by superior solutions. For every $\mathbf{a} \in \mathcal{A}$, let

$$B_S(\mathbf{a}) = \{\mathbf{a}' \in \mathcal{A} : v_{\mathcal{R}_n}(\mathbf{a}') < v_{\mathcal{R}_n}(\mathbf{a}) \text{ and } v_{\mathcal{S}_n}(\mathbf{a}') \geq v_{\mathcal{S}_n}(\mathbf{a})\}$$

$$B_R(\mathbf{a}) = \{\mathbf{a}'_n \in \mathcal{A}_n : v_{\mathcal{R}_n}(\mathbf{a}') \leq v_{\mathcal{R}_n}(\mathbf{a}) \text{ and } v_{\mathcal{S}_n}(\mathbf{a}') > v_{\mathcal{S}_n}(\mathbf{a})\},$$

and define the set of alternatives that are *strictly better* than $\mathbf{a}$ as follows:

$$B(\mathbf{a}) = B_S(\mathbf{a}) \cup B_R(\mathbf{a}).$$

The set $B(\mathbf{a})$ consists of all possible alternatives that are either less rejectable and not less selectable than $\mathbf{a}$, or more selectable and not more rejectable than $\mathbf{a}$. If $B(\mathbf{a}) = \varnothing$, then no alternative can be preferred to $\mathbf{a}$ in both selectability and rejectability.

Definition 6.11 The *nondominated satisficing* set is $\mathfrak{N}_{\mathcal{X}_n} = \{\mathbf{a} \in \mathcal{A} : B(\mathbf{a}) = \varnothing\}$, and the *satisficing compromise set* is $\mathfrak{C}_{\mathcal{X}_n} = \mathfrak{N}_{\mathcal{X}_n} \cap \mathbf{C}_{\mathcal{X}_n}$. □

All elements of $\mathfrak{C}_{\mathcal{X}_n}$ lay claim to some notion of optimality. For example,

$$\mathbf{a}^* = \arg \max_{\mathbf{a} \in \mathfrak{C}_{\mathcal{X}_n}} \{v_{\mathbf{S}_n}(\mathbf{a}) - q_{\mathcal{X}_n} v_{\mathbf{R}_n}(\mathbf{a})\}$$

maximizes the difference between selectability and rejectability;

$$\mathbf{a}^{**} = \arg \min_{\mathbf{a} \in \mathfrak{C}_{\mathcal{X}_n}} v_{R_n}(\mathbf{a})$$

minimizes rejectability, and

$$\mathbf{a}^{***} = \arg \max_{\mathbf{a} \in \mathfrak{C}_{\mathcal{X}_n}} v_{\mathbf{R}_n}(\mathbf{a})$$

maximizes selectability. The remaining elements of $\mathfrak{C}_{\mathcal{X}_n}$ represent other optimal trade-offs between effectiveness and efficiency. As $q \to 0$, it is clear that an optimal compromise will eventually exist.

6.6 Satisficing social choice

Section 3.9 introduces the conditional social choice problem, which we now express in the satisficing context. Recall that the key difference between game theory and social choice theory is that, with the latter, the preferences of the individuals are aggregated to form a choice for the group; whereas with the former, each individual is empowered to implement its own choice. Thus, the alternative set $\mathcal{A}$ is common to all stakeholders.

Definition 6.12 A *conditional satisficing social choice problem* is a quadruple $\{\mathcal{S}_n, \mathcal{R}_n, \mathcal{A}, W_{\mathcal{S}_n \mathcal{R}_n}\}$, where the social choice concordant utility $W_{\mathcal{S}_n \mathcal{R}_n}$ (see (3.7) and (6.6)) collapses to the form

$$W_{\mathcal{S}_n \mathcal{R}_n}(s_1, \ldots, s_n, r_1, \ldots, r_n) = \prod_{i=1}^{n} u_{S_i | \mathrm{pa}(S_i)}(s_i | s_{k_1}, \ldots, s_{k_{p_i}}, r_{m_1}, \ldots, r_{m_{z_i}})$$

$$\prod_{j=1}^{n} u_{R_j | \mathrm{pa}(R_j)}(r_j | s_{l_1}, \ldots, s_{l_{t_j}}, r_{l_1}, \ldots, r_{l_{v_j}}),$$

$$\tag{6.11}$$

where the profiles $(s_1, \ldots, s_n)$ and $(r_1, \ldots, r_n)$ are joint social choice conjectures for the selecting and rejecting selves, respectively. Notice that the satisficing group welfare function $v_{\mathcal{S}_n \mathcal{R}_n}$ coincides with the concordant utility. $\square$

The *selectability concordance utility* and the *rejectability concordance utility* are given by

$$W_{\mathcal{S}_n}(s_1, \ldots, s_n) = \sum_{r_1, \ldots, r_n} W_{\mathcal{S}_n \mathcal{R}_n}(s_1, \ldots, s_n, r_1, \ldots, r_n)$$

and

$$W_{\mathcal{R}_n}(r_1, \ldots, r_n) = \sum_{s_1, \ldots, s_n} W_{\mathcal{S}_n \mathcal{R}_n}(s_1, \ldots, s_n, r_1, \ldots, r_n),$$

respectively.

The *satisficing social welfare function* is defined as

$$w_{\mathcal{X}_n}(a) = W_{\mathcal{S}_n}(a, \ldots, a) - q_{\mathcal{X}_n} W_{\mathcal{R}_n}(a, \ldots, a), \tag{6.12}$$

where $q_{\mathcal{X}_n}$ is the group-level boldness index.

The *individual selectability welfare function* for S_i is

$$w_{S_i}(a_i) = \sum_{\sim a_i} W_{\mathcal{S}_n}(a_1, \ldots, a_n),$$

and the *individual rejectability welfare function* for R_i is

$$w_{R_i}(a_i) = \sum_{\sim a_i} W_{\mathcal{R}_n}(a_1, \ldots, a_n).$$

The *jointly satisficing social choice set* is

$$\boldsymbol{\Sigma}_{\mathcal{X}_n} = \{a \in \mathcal{A} \colon w_{\mathcal{X}_n}(a) \geq 0\},$$

and the *individual satisficing social choice set* for X_i is

$$\Sigma^i_{q_{X_i}} = \{a \in \mathcal{A} \colon w_{S_i}(a) \geq q_{X_i} w_{R_i}(a)\}. \tag{6.13}$$

The *simultaneously satisficing social choice set* is

$$\sigma_{\mathcal{X}_n} = \Sigma^1_{q_{X_1}} \cap \cdots \cap \Sigma^n_{q_{X_n}}, \tag{6.14}$$

and the *satisficing social compromise set* is

$$\mathcal{C}_{\mathcal{X}_n} = \sigma_{\mathcal{X}_n} \cap \boldsymbol{\Sigma}_{\mathcal{X}_n}.$$

The *satisficing social compromise* is the action in this set that maximizes the social welfare within the compromise set; that is,

$$a^* = \arg \max_{a \in C_{\mathcal{X}_n}} w_{\mathcal{X}_n}(a).$$

If $\Sigma_{\mathcal{X}_n} = \varnothing$, there is no alternative that is good enough for the group; if $\sigma_{\mathcal{X}_n} = \varnothing$, there is no alternative that is simultaneously good enough for all individuals; and if $C_{\mathcal{S}_n} = \varnothing$, there is no alternative that is simultaneously good enough for all individuals *and* the group. However, by reducing the q-values incrementally, a compromise will eventually emerge.

6.7 Coordination

There are a number of possible ways to investigate coordination in the satisficing context: (a) the interself coordination between selecting and rejecting subgroups, (b) the intraself coordination within the members of the selecting subgroup and within the members of the rejecting subgroup, and (c) coordination between the members of the group without distinctions among the selves.

6.7.1 Interself coordination

To investigate the coordination capacity between selecting and rejecting subgroups, let $\mathcal{X}_k$, $k \le n$ be an arbitrary subgroup of $\mathcal{X}_n$. We must compute the joint selecting/rejecting hardness, yielding $H(\mathcal{S}_k, \mathcal{R}_k) = H(S_{i_1}, \ldots, S_{i_k}, R_{i_1}, \ldots, R_{i_k})$, the joint selecting hardness $H(\mathcal{S}_k) = H(S_{i_1}, \ldots, S_{i_k})$, and the joint rejecting hardness $H(\mathcal{R}_k) = H(R_{i_1}, \ldots, R_{i_k})$ as follows:

$$H(\mathcal{S}_k, \mathcal{R}_k) = -\sum_{\mathbf{s}} \sum_{\mathbf{r}} v_{\mathcal{S}_k \mathcal{R}_k}(\mathbf{s}, \mathbf{r}) \log v_{\mathcal{S}_k \mathcal{R}_k}(\mathbf{s}, \mathbf{r}),$$

$$H(\mathcal{S}_k) = -\sum_{\mathbf{a}} v_{\mathcal{S}_k}(\mathbf{a}) \log v_{\mathcal{S}_k}(\mathbf{a}),$$

and

$$H(\mathcal{R}_k) = -\sum_{\mathbf{a}} v_{\mathcal{R}_k}(\mathbf{a}) \log v_{\mathcal{R}_k}(\mathbf{a}).$$

The interself mutual information between the selecting and rejecting subgroups is

$$I_{interself}(\mathcal{S}_k, \mathcal{R}_k) = H(\mathcal{S}_k) + H(\mathcal{R}_k) - H(\mathcal{S}_k, \mathcal{R}_k),$$

the corresponding interself distance metric is

$$d_{interself}(\mathbf{S}_k, \mathbf{R}_k) = (2k-1)H(\mathbf{S}_k, \mathbf{R}_k) - I_{interself}(\mathbf{S}_k, \mathbf{R}_k),$$

the interself relative dispersion measure is

$$\mathcal{D}_{interself}(\mathbf{S}_k, \mathbf{R}_k) = \frac{1}{2k-1} \frac{d_{interself}(\mathbf{S}_k, \mathbf{R}_k)}{H(\mathbf{S}_k, \mathbf{R}_k)},$$

and the interself coordination capacity is

$$C_{interself}(\mathbf{S}_k, \mathbf{R}_k) = 1 - \mathcal{D}_{interself}(\mathbf{S}_k, \mathbf{R}_k). \tag{6.15}$$

Equation (6.15) provides a measure of the ability of the selecting subgroup $\mathbf{S}_k$ to coordinate with the rejecting subgroup $\mathbf{R}_k$, and thus is a measure of the benefit/cost tension inherent in the subgroup $\mathbf{X}_k$.

6.7.2 Intraself coordination

We may also consider the ability of the selecting selves to coordinate with themselves as a subgroup, similarly, for the rejecting selves. To proceed, we first compute the individual hardness for each self, yielding

$$H(S_i) = -\sum_{a_i} v_{S_i}(a_i) \log v_{S_i}(a_i)$$

$$H(R_i) = -\sum_{a_i} v_{R_i}(a_i) \log v_{R_i}(a_i).$$

We next account for the intraself selecting and rejecting mutual information functions as

$$I_{intraself}(\mathbf{S}_n) = \sum_{i=1}^{n} H(S_i) - H(\mathbf{S}_n)$$

$$I_{intraself}(\mathbf{R}_n) = \sum_{i=1}^{n} H(R_i) - H(\mathbf{R}_n).$$

The corresponding intraself distance measures are

$$d_{intraself}(\mathbf{S}_n) = (n-1)H(\mathbf{S}_n) - I_{intraself}(\mathbf{S}_n)$$

$$d_{intraself}(\mathbf{R}_n) = (n-1)H(\mathbf{R}_n) - I_{intraself}(\mathbf{R}_n),$$

the intraself selecting relative dispersion measures are

$$\mathcal{D}_{intraself}(\boldsymbol{\mathcal{S}}_n) = \frac{1}{n-1} \frac{d_{intraself}(\boldsymbol{\mathcal{S}}_n)}{H(\boldsymbol{\mathcal{S}}_n)}$$

$$\mathcal{D}_{intraself}(\boldsymbol{\mathcal{R}}_n) = \frac{1}{n-1} \frac{d_{intraself}(\boldsymbol{\mathcal{R}}_n)}{H(\boldsymbol{\mathcal{R}}_n)}$$

and the intraself selecting and rejecting coordination indices are

$$C_{intraself}(\boldsymbol{\mathcal{S}}_n) = 1 - \mathcal{D}_{intraself}(\boldsymbol{\mathcal{S}}_n)$$

$$C_{(intraself}\boldsymbol{\mathcal{R}}_n) = 1 - \mathcal{D}_{intraself}(\boldsymbol{\mathcal{R}}_n).$$

6.7.3 Interagent coordination

Perhaps the most meaningful notion of coordination is the ability of subgroups of agents to coordinate with each other. We first consider the pairwise coordination between X_i and X_j. To proceed, we must compute the joint social welfare of the two stakeholders as

$$v_{S_i S_j R_i R_j}(s_i, s_j, r_i, r_j) = \sum_{\sim (s_i, s_j, r_i, r_j)} v_{\boldsymbol{\mathcal{S}}_n \boldsymbol{\mathcal{R}}_n}(s_1, \ldots, s_n, r_1, \ldots, r_n),$$

from which we compute the hardness

$$H(S_i, R_i, S_j, R_j) = - \sum_{s_i, s_j, r_i, r_j} v_{S_i S_j R_i R_j}(s_i, s_j, r_i, r_j) \log v_{S_i S_j R_i R_j}(s_i, s_j, r_i, r_j).$$

We next compute the marginals

$$v_{S_i R_i}(s_i, r_i) = \sum_{\sim (s_j, r_j)} v_{S_i S_j R_i R_j}(s_i, s_j, r_i, r_j)$$

$$v_{S_j R_j}(s_i, r_i) = \sum_{\sim (s_j, r_j)} v_{S_i S_j R_i R_j}(s_i, s_j, r_i, r_j),$$

from which we compute the agent-level hardness values

$$H(S_i, R_i) = - \sum_{(s_i, r_i)} v_{S_i R_i}(s_i, r_i) \log v_{S_i R_i}(s_i, r_i)$$

$$H(S_j, R_j) = - \sum_{(s_j, r_j)} v_{S_j R_j}(s_j, r_j) \log v_{S_j R_j}(s_j, r_j).$$

The interagent mutual information is

$$I_{interagent}(S_i, R_i, S_j, R_j) = H(S_i, R_i) + H(S_j, R_j) - H(S_i, S_j, R_i, R_j),$$

the interagent distance measure is

$$d_{interagent}(S_i, S_j, R_i, R_j) = 3H(S_i, S_j, R_i, R_j) - I_{interagent}(S_i, R_i, S_j, R_j),$$

$$\mathcal{D}_{interagent}(S_i, R_i, S_j, R_j) = \frac{d_{interagent}(S_i, S_j, R_i, R_j)}{2H(S_i, S_j, R_i, R_j)},$$

and, finally,

$$C_{interagent}(X_i, X_j) = C_{interagent}(S_i, R_i, S_j, R_j) = 1 - \mathcal{D}_{interagent}(S_i, R_i, S_j, R_j).$$

6.7.4 Intragroup coordination

We now consider the coordination capacity of the entire $2n$-member group without making distinctions among the selves. The mutual information is

$$I_{intragroup}(S_1, \ldots S_n, R_1, \ldots, R_n)$$
$$= \sum_i H(S_i) + \sum_i H(R_i) - H(S_1, \ldots, S_n, R_1, \ldots, R_n),$$

the distance measure is

$$d_{intragroup}(S_1, \ldots, S_n, R_1, \ldots, R_n) = (2n - 1)H(S_1, \ldots, S_n, R_1, \ldots, R_n)$$
$$- I_{intragroup}(S_1, \ldots S_n, R_1, \ldots, R_n),$$

the relative dispersion measure is

$$\mathcal{D}_{intragroup}(S_1 \ldots, S_n, R_1, \ldots, R_n) = \frac{1}{2n - 1} \frac{d_{intragroup}(S_1, \ldots, S_n, R_1, \ldots, R_n)}{H(S_1, \ldots, S_n, R_1, \ldots, R_n)},$$

and the interself coordination capacity is

$$C_{intragroup}(S_1, \ldots, S_n, R_1, \ldots, R_n) = 1 - \mathcal{D}_{intragroup}(S_1 \ldots, S_n, R_1, \ldots, R_n).$$

6.8 Summary

The concept of satisficing discussed in this chapter, although somewhat differ-ent from the usage advocated by Simon (1955), is designed according to the same fundamental principle, namely, to identify solutions that are good enough. The main philosophical distinction between the two concepts is that, whereas satisficing as used by Simon defines the notion of "good enough" heuristically, our approach defines it in terms of a benefit-to-cost comparison. Satisficing as used by Simon is an approximation to the conventional optimizing dictum to settle for the best and only the best solution. In contrast, satisficing as used here is more along the lines of getting your money's worth or getting what you pay for.

Satisficing game theory is an extension of conditional game theory as developed in Chapters 2 and 3. By decomposing each stakeholder into a selecting self and a rejecting self, a satisficing game allows a more finely tuned specification of each stakeholder's preferences in terms of effectiveness and efficiency. As with conditional games, satisficing games permit the use of conditional utilities, thereby permitting a more careful characterization of the social influence relationships among the selves. Also, as with conditional games, satisficing games permit solutions that remove the wedge between group-level interests and individual-level interests, thereby enabling the simultaneous consideration of both levels of interest.

Several major benefits devolve from the satisficing notion. The first is the introduction of two operationally distinct selves for each stakeholder: one that is concerned with the effectiveness of achieving the goal of the stakeholder independently of efficiency considerations, and one that is concerned with the costs of taking action independently of its effectiveness. The second benefit is that the satisficing theory readily extends to the multistakeholder case, where, by applying the conditional utility syntax, we extend the spheres of interest beyond self-interest, thereby accommodating social relationships. The third major benefit of the satisficing approach to decision making is that it is explicitly designed to provide a set of "good enough" solutions, rather than a single "best" solution. This capability provides stakeholders increased flexibility to choose the action they ultimately wish to implement. In a way that is somewhat similar to the Pareto frontier, the satisficing set provides the set of all solutions that are not dominated according to their selecting and rejecting attributes.

7

Applications

Nothing is more practical than a good theory.

— Ludwig Boltzmann

In this chapter we present some application examples that are more lengthy than the examples discussed in earlier chapters. Although these examples are heavily stylized, they illustrate essential features of decision making in various contexts. The first three examples, the Battle of the Sexes, Ultimatum, and the Stag Hunt games, are games, along with the Prisoner's Dilemma game discussed in earlier chapters, that have received a great deal of attention as important examples of game theory, since they serve as models of many real-world situations. The fourth example, the Family Walk, is a social choice example that illustrates how conditional utilities can be used to define a social network in which, although every participant gets a vote, the votes are not delivered in a social vacuum. The fifth and final example is that of a multiagent system comprising three autonomous decision makers that must function cooperatively.

7.1 Battle of the Sexes

As discussed in Example 1.3, the Battle of the Sexes game involves a man and a woman who plan to meet in town for a social function. She (S) prefers to go to the ballet (B), whereas he (H) prefers the dog races (D). Each prefers to be with the other, however, regardless of where the social function takes place. This game serves as a model for economic scenarios where two firms choose between competing standards. Although each has its own preference, both would sell more products if they were to adopt a common standard. The payoff matrix in ordinal form is given in Table 1.3. This game has two Nash equilibria, (D, D) and (B, B). Both are also Pareto optimal, but there is no dominant strategy. Notice that if both players were categorically altruistic, H would go to B in an attempt to please S, whereas S would go to D in an attempt to please H, resulting in the worst outcome for both.

This game illustrates the shortcomings of classical theory for the characterization of behavior when cooperation is important. Each player's utility is a measure of the individual's own enjoyment, regardless of the enjoyment of the other. The strategy vector (D, D) is best for H, but it is because H gets his way on both counts: He goes to his favorite event and is with S. S's preferences for the venue, however, are not taken into consideration by H. According to this framing, it would not matter to H if S detested dog races but were willing to put up with that event at great sacrifice of enjoyment just to be with H. The problem is that optimization fosters a subtle form of competition, even though cooperation and even deference may more naturally occur.

To construct a satisficing framing of this game, we must first establish each player's notions of failure avoidance and resource conservation. In the interest of simplifying our development, we shall assume that the players' utilities are completely dissociated. We associate failure avoidance with the most important goal of the game, which is for the two players to be with each other, regardless of the venue. Resource conservation, on the other hand, deals with the costs of being at a particular function. Obviously, H would prefer D if he did not take into consideration S's preferences; similarly, S would prefer B. Thus, we express the rejectabilities for H and S in terms of parameters h and s, respectively, as

$$u_{R_H}(D) = h$$
$$u_{R_H}(B) = 1 - h$$

and

$$u_{R_S}(D) = 1 - s$$
$$u_{R_S}(B) = s \,,$$

where h is H's rejectability of D and s is S's rejectability of B. The closer h is to zero, the more H is adverse to B with an analogous interpretation for s with respect to S attending D. To be consistent with the stereotypical roles, we may assume that $0 \leq h < \frac{1}{2}$ and $0 \leq s < \frac{1}{2}$. As will be subsequently seen, only the ordinal relationship need be specified, that is, either $s < h$ or $h < s$.

Selectability is a measure of the failure avoidance associated with the alternatives. Since being together is a joint, rather than an individual, objective, it is difficult to form unilateral assessments of selectability, but it is possible to characterize individually the conditional selectability. To do so requires the specification of the conditional selectability functions $u_{S_H|R_S}$ and $u_{S_S|R_H}$; that is, H's selectability conditioned on S's rejectability and S's selectability conditioned on H's rejectability. If S were to place her entire unit mass of rejectability on D, H may account for this, if he cares at all about S's feelings, by placing some portion of his conditional selectability mass on B. S may construct her conditional selectability in a similar way, yielding

$$u_{S_H|R_S}(D|D) = 1 - \alpha$$

$$u_{S_H|R_S}(B|D) = \alpha$$

$$u_{S_H|R_S}(D|B) = 1$$

$$u_{S_H|R_S}(B|B) = 0$$

and

$$u_{S_S|R_H}(D|D) = 0$$

$$u_{S_S|R_H}(B|D) = 1$$

$$u_{S_S|R_H}(D|B) = \beta$$

$$u_{S_S|R_H}(B|B) = 1 - \beta.$$

Figure 7.1 illustrates the utility network corresponding to this game.

The valuations $u_{S_H|R_S}(B|D) = \alpha$ and $u_{S_S|R_H}(D|B) = \beta$ can be considered conditions of conditional altruism. If S were to place all of her rejectability mass on D, then H may defer to S's strong dislike of D by placing α of his selectability mass, *as conditioned by her preference,* on B. Similarly, S could show a symmetrical conditional preference for D if H were to reject B strongly. The parameters α and β are H's and S's *indices of conditional altruism,* respectively, and serve as a way for each to control the amount of deference he or she is willing to grant to the other. In the interest of simplicity, we shall assume that both players are maximally altruistic and set $\alpha = \beta = 1$. In

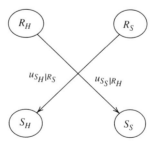

Figure 7.1. The utility network for the satisficing Battle of the Sexes game.

Table 7.1. *The satisficing group welfare function* $v_{S_H S_S R_H R_S}(s_1, s_2, r_1, r_2)$ *for the satisficing Battle of the Sexes game*

(s_H, s_S)	(D,D)	(D,B)	(B,D)	(B,B)
			(r_H, r_S)	
(D,D)	0	0	0	$(1-h)s$
(D,B)	0	hs	0	0
(B,D)	0	0	$(1-h)(1-s)$	0
(B,B)	$h(1-s)$	0	0	0

principle, however, the indices may be set independently to any value in $[0,1]$. Notice that even in the most altruistic case, the conditional preferences do not commit one to categorical abdication of his or her own unilateral preferences. H may still myopically (that is, without taking S into consideration) prefer D, and S still myopically prefer B, and there is no intimation that either participant must throw the game in order to accommodate the other.

With these conditional and marginal functions, we may construct the satisficing group welfare function (which is coincident with the concordant utility under the complete dissociation assumption) as

$$v_{S_H S_S R_H R_S}(s_1, s_2, r_1, r_2) = u_{S_H | R_S}(s_1 | r_2) u_{S_S | R_H}(s_2 | r_1) u_{R_H}(r_1) u_{R_S}(r_2), \quad (7.1)$$

where we assume that H's selectability conditioned on S's rejectability is dependent only on S's rejectability, that S's selectability conditioned on H's rejectability is dependent only on H's rejectability, and that the myopic rejectability values of H and S are independent. Table 7.1 displays the values assumed by this satisficing group welfare function.

Application of (6.8) and (6.9) results in joint selectability and rejectability values of

$$v_{S_H S_S}(D,D) = (1-h)s$$
$$v_{S_H S_S}(D,B) = hs$$
$$v_{S_H S_S}(B,D) = (1-h)(1-s)$$
$$v_{S_H S_S}(B,B) = h(1-s)$$

and

$$v_{R_H R_S}(D,D) = h(1-s)$$
$$v_{R_H R_S}(D,B) = hs$$

$$v_{R_H R_S}(B, D) = (1 - h)(1 - s)$$

$$v_{R_H R_S}(B, B) = (1 - h)s.$$

The marginal selectability and rejectability values for H and S are

$$v_{S_H}(D) = s \qquad\qquad v_{R_H}(D) = h$$

$$v_{S_H}(B) = 1 - s \qquad\qquad v_{R_H}(B) = 1 - h$$

and

$$v_{S_S}(D) = 1 - h \qquad\qquad v_{R_S}(D) = 1 - s$$

$$v_{S_S}(B) = h \qquad\qquad v_{R_S}(B) = s.$$

Setting the index of boldness, q, equal to unity, the jointly satisficing set is

$$\Sigma_q = \begin{cases} \{(D,B),(B,D),(B,B)\} & \text{for } s < h \\ \{(D,D),(D,B),(B,D)\} & \text{for } s > h \quad, \\ \{(D,D),(D,B),(B,D),(B,B)\} & \text{for } s = h \end{cases}$$

the individually satisficing sets are

$$\Sigma_q^H = \begin{cases} \{B\} & \text{for } s < h \\ \{D\} & \text{for } s > h \\ \{B,D\} & \text{for } s = h \end{cases}$$

$$\Sigma_q^S = \begin{cases} \{B\} & \text{for } s < h \\ \{D\} & \text{for } s > h \quad, \\ \{B,D\} & \text{for } s = h \end{cases}$$

and the jointly satisficing rectangle is

$$\mathcal{R}_{HS} = \Sigma_q^H \times \Sigma_q^S = \begin{cases} \{B,B\} & \text{for } s < h \\ \{D,D\} & \text{for } s > h \quad. \\ \{\{B,B\},\{D,D\}\} & \text{for } s = h \end{cases}$$

By inspection, we see that the compromise set is identical to the satisficing rectangle for this problem. Thus, if S's aversion to D is less than H's aversion to B, then both players will go to H's preference, namely, D, and vice versa. Consequently, even though both players are maximally conditionally deferential, the satisficing solution results in a very natural cooperative strategy.

We now examine the coordination capacity of the Battle of the Sexes game. We first compute the marginals for each agent, yielding, from (7.1),

$$v_{S_H R_H}(s_H, r_H) = \sum_{\sim(s_S, r_S)} v_{S_H S_S R_H R_S}(s_H, s_S, r_H, r_S)$$

$$v_{S_S R_S}(s_S, r_S) = \sum_{\sim(s_H, r_H)} v_{S_H S_S R_H R_S}(s_H, s_S, r_H, r_S),$$

which becomes

$$v_{S_H R_H}(D, D) = hs$$
$$v_{S_H R_H}(D, B) = (1 - h)s$$
$$v_{S_H R_H}(B, D) = h(1 - s)$$
$$v_{S_H R_H}(B, B) = (1 - h)(1 - s)$$

and

$$v_{S_S R_S}(D, D) = (1 - h)(1 - s)$$
$$v_{S_S R_S}(D, B) = (1 - h)s$$
$$v_{S_S R_S}(B, D) = h(1 - s)$$
$$v_{S_S R_S}(B, B) = hs.$$

Direct computation yields

$$H(S_H, S_S, R_H, R_S) = -(1 - h)s \log(1 - h)s - hs \log hs - h(1 - s) \log h(1 - s)$$
$$- (1 - h)(1 - s) \log(1 - h)(1 - s).$$

Furthermore, we see by inspection that

$$H(S_H, R_H) = H(S_S, R_S) = H(S_H, S_S, R_H, R_S).$$

Also,

$$I_{interagent}(S_H, R_H, S_S, R_S) = H(S_H, R_H) + H(S_S, R_S)$$
$$- H(S_H, S_S, R_H, R_S) = H(S_H, R_H)$$

and, consequently,

$$d_{interagent}(S_H, R_H, S_S, R_S) = 2H(S_H, R_H)$$

and, finally, $\mathcal{D}_{interagent}(S_H, R_H, S_S, R_S) = \frac{2}{3}$, so

$$C_{interagent}(X_H, X_S) = C_{interagent}(S_H, R_H, S_S, R_S) = \frac{1}{3}.$$

Thus, we see that the coordination capacity of the satisficing Battle of the Sexes game is independent of the values of s and h. The reason for this independence is that it is only the ordinal relationship between s and h, and not the cardinal relationship, that determines the satisficing solution. The value of one-third is an indication that the participants in this game possess a modest but not negligible intrinsic ability to coordinate their behaviors and function harmoniously.

7.2 The Ultimatum game

The Ultimatum game has received great attention as a purported example of irrational behavior; that is, as a case where the players of the game are motivated by considerations other than maximizing expected utility. Players often do not act as predicted by the principle of individual rationality, according to which the players would act dispassionately according to the rankings of their payoffs. Instead, people often show a bias toward fair play, which is an indication that they are motivated by psychological factors that are not encoded in the payoffs. Two explanations are possible: Either they are altering their utilities from the payoffs in an attempt to account for the social environment but remaining true to individual rationality, or they are abandoning narrow self-interest for some other notion of rational behavior. In either case, conventional game theory cannot easily accommodate such deviant behavior of the players – hence the apparent irrational behavior.

The setup of this two-player game was described in Example 1.2 but is repeated here for completeness. X_1, called the *proposer*, offers X_2, called the *responder*, a fraction (chosen by X_1) of a fortune, and X_2 chooses whether or not to accept the offer. If X_2 accepts, then the two players divide the fortune between themselves according to the agreed-upon ratio. If X_2 declines the offer, neither player receives anything. In both cases, the game is over; there is no opportunity for reconsideration. The Nash equilibrium solution to this game is for X_1 to offer the smallest nonzero amount it can, and for X_2 to accept whatever is offered. This is the play predicted by individual rationality. Interestingly, such a strategy is rarely adopted by human players. Even with one-off play, proposers are inclined to offer fair deals, and responders are inclined to reject unfair deals (Binmore, 1998; Nowak et al., 2000; Sigmund et al., 2002).

There have been many attempts to explain this phenomenon, with the most popular explanations being the hypotheses that (a) humans find it difficult to permit a rival to gain an advantage, and (b) evolutionary conditioning of group dependency motivates players to maintain socially acceptable reputations. According to these explanations, the players have modified their utilities to account for social considerations such as jealousy or disrespect. It is assumed, however, that their notion of rationality has not changed. They are still playing to maximize their expected utility within the context of classical game theory; those utilities just happen to be other than the financial payoff values.

Game theory draws much of its strength from the practice of abstracting the essential features of interagent relationships into a payoff array that can then be analyzed on the basis of individual rationality to yield normative solution concepts such as Nash equilibria. The apparent need with the Ultimatum game to modify the story line to explain behavior that does not conform to a priori expectations is a manifestation of either (a) an inadequate game-theoretic model for the situation or (b) the fundamental limitations of classical game theory to deal with the situation.[1] Most discussions of this game that appear in the literature seem to favor the former explanation and attempt to append extra-game-theoretic considerations, such as the emotional state of the players, to justify the apparently irrational behavior. But perhaps an equally plausible source of the limitations lies with the structure of the utilities.

Two issues are relevant here. First, if sociological attributes are material considerations for the behavior of the players of a game, it seems reasonable that such attributes should be embedded explicitly into the utilities, rather than being merely appended as post factum explanations for behavior. Perhaps too much has been abstracted from the story line in the formulation of the game to ensure that a payoff array captures all of the essential attributes of the players.

The second issue is that, taking the results of the Ultimatum game at face value, it is difficult to cling to the hypothesis that people are utility maximizers. There is much empirical evidence to the contrary, and it is at least an open question. Consequently, reasonable alternative hypotheses deserve to be investigated. In this section we show how conditional game theory and, in particular, the satisficing variant, can be used to characterize the Ultimatum game such that (a) relevant sociological considerations appear as explicit components of the game setup, and (b) the demand for utility maximization can be replaced by a less-rigid notion of acceptability.

Under its original formulation, the Ultimatum game permits X_1 to offer any fraction of the fortune to X_2, and thus X_1 has a continuum of options at its disposal, whereas X_2 has only two options: a (accept) or r (reject). The game loses little of its effect, and its analysis is much simpler if we follow the lead of Gale et al. (1995) and   consider the so-called minigame, with only two alternatives for the proposer: h and l (high and low), with $0 < l < h \leq \frac{1}{2}$. These

---

[1] For many game theorists, the game *is* the payoff matrix; the story line is incidental. For example, the story line that gives the Prisoner's Dilemma its name was invented post factum to conform, for pedagogical illustration, to the payoff matrix with the given structure, rather than the other way around.

Table 7.2. *The payoff matrix for the*
Ultimatum minigame

	X_2	
X_1	a	r
h	$(1-h,\, h)$	$(0,\, 0)$
l	$(1-l,\, l)$	$(0,\, 0)$

values correspond to the fraction of the fortune that X_1 is prepared to offer to X_2. This minigame analysis captures the essential features of the continuum game and permits us to see more clearly the relationships between the two players. With this restriction, the action sets for the two players are $A_1 = \{h, l\}$ and $A_2 = \{a, r\}$ for X_1 and X_2, respectively.

The payoff matrix for the Ultimatum minigame is illustrated in Table 7.2. This game has a dominant strategy for each player; namely, X_1 should play l and X_2 should play a. The joint strategy $\{l, a\}$ is also the unique Nash equilibrium, and according to the doctrine of individual rationality, the players should adopt this joint strategy. The response of many players of this game, however, is an indication that there is more to this game than meets the eye as viewed through the payoff matrix. In particular, players of this game are typically *not* utility maximizers – at least, that is, if utility is expressed in terms of the fortune that is to be divided. Analysts of the game have theorized that the responders decline an offer they deem to be unfair because they are emotionally connected with the consequence. They may be jealous of the proposer, they may be resentful of being scorned, or they may be protective of their reputation. A term that seems to capture these, as well as possibly other emotional responses, is the term *indignation*. If the responder's indignation is relevant to the game, a parameter representing this attribute should be part of the mathematical structure of the game.

Another feature that emerges from the play of this game is the possibility that the proposer may be motivated by considerations other than greed. Altruism is certainly a possibility, but there is also the pragmatic notion that the responder may reject the offer if the proposer appears to be excessively greedy. Even if the proposer is greedy, it may still make an equitable offer if it suspects that the responder would be prone to reject an inequitable one; but if the proposer were not so prone, it would be inclined to make a inequitable offer. A concept that expresses this emotional attribute is *intemperance*. Thus, we shall take the intemperance of the proposer and the indignation of the responder as the dominant social attributes of the players. We will denote these two attributes by the intemperance index τ and the indignation index δ and assume that $0 \leq \tau \leq 1$ and $0 \leq \delta \leq 1$. The condition $\tau \approx 1$ means that the proposer is extremely avaricious, whereas $\tau \approx \frac{1}{2}$ means that it is moderate and willing to restrain its demand for wealth. If $\tau < \frac{1}{2}$, then the proposer may be viewed as having altruistic tendencies. The condition $\delta \approx 1$ means that the responder is easily offended, whereas $\delta \approx 0$ means that it is extremely tolerant. Although it may be reasonable to assume that these social parameters are somehow tied to the fractions h and l, we do not rely on any such assumptions. For purposes of our analysis, these parameters are fixed properties of the players (they may be learned by repeated play, but that discussion is beyond the scope of this book).

Once we have models for the players' behaviors in terms of both monetary and social attributes, the next step is to parse the players into their selecting and rejecting selves. There is not a unique way to define these selves, but it generally makes sense to associate the selecting self with the fundamental goal, which, for this game, is to receive as much reward as possible. It also makes sense to associate the rejecting self with the negative attribute of a decision, which, for this game, is the complete forfeiture of the reward. Regardless of the way the selves are defined, however, it is essential that the selecting and rejecting attributes of an option are not simply restatements of the same attribute.

There are four selves for this game: S_1, S_2, R_1, and R_2, and once we have defined the relationships among them, we will have a complete characterization of the game from both the monetary and social perspectives.

A significant feature of the Ultimatum game is that the players do not make simultaneous decisions. Since the proposer (X_1) goes first, its utility structure, although it is governed by X_2's possible responses, is not *conditioned* on X_2's response (with games involving simultaneous moves, each player may condition its actions on the anticipated choices of the other). Thus, X_1's utility is categorical. X_2, however, must respond to X_1's decision; thus a conditional utility is appropriate.

We first consider the proposer's utilities. The general form of these utilities is $u_{S_1}(s_{11},s_{12})$ and $u_{R_1}(r_{11},r_{12})$. A reasonable simplification of this problem is to assume a condition of complete dissociation, resulting in $u_{S_1}(s_1)$ and $u_{R_1}(r_1)$ (we no longer need the double-subscript notation). The proposer's goal of achieving the larger fraction of the fortune is modulated only by its temerity. Thus,

$$u_{S_1}(h) = 1 - \tau \quad \text{and} \quad u_{S_1}(l) = \tau . \tag{7.2}$$

For $\tau > \frac{1}{2}$, the proposer prefers the higher reward and so prefers to offer the responder the low fraction.

In the interest of avoiding forfeiture, the proposer would be wise to modulate its temerity by considering the indignation that the responder would feel if a low offer were tendered, yielding

$$u_{R_1}(h) = \tau(1-\delta) \quad \text{and} \quad u_{R_1}(l) = 1 - \tau(1-\delta);\tag{7.3}$$

that is, if the responder were highly indignant ($\delta \approx 1$), then the proposer would ascribe low rejectability to offering the high fraction to the responder.

Next, we consider the responder's utilities, which we also assume to be completely dissociated. Since X_2 makes the second move, the preferences of its selves will be conditioned on X_1's choice. Let us first consider S_2's response to S_1. Our goal is to form the conditional utility $u_{S_2|S_1}$, which characterizes how X_2 would view its options in terms of achieving its goal without taking into consideration the possible negative consequences, given that X_1 were to make its choice also without taking into consideration the negative consequences of that choice. In keeping with our monetary and social models of the Ultimatum game, we define this conditional utility as follows:

- If S_1 were to select h, then S_2 would select a. Thus, we set

$$u_{S_2|S_1}(a|h) = 1 \quad \text{and} \quad u_{S_2|S_2}(r|h) = 0.$$

- If S_1 were to select l, then S_2 would be indignant and would select r with weight δ and a with weight $1 - \delta$. Thus, we set

$$u_{S_2|S_1}(a|l) = 1 - \delta \quad \text{and} \quad u_{S_2|S_1}(r|l) = \delta.$$

Next, let us consider how X_2's rejecting self, R_2, would respond to X_1's selecting self.

- If S_1 were to select h, then S_2 would certainly accept that portion and would reject r. Thus, we set

$$u_{R_2|S_1}(a|h) = 0 \quad \text{and} \quad u_{R_2|S_1}(r|h) = 1.$$

- If S_1 were to select l, then S_2 would be indignant and would reject a with weight δ and r with weight $1 - \delta$. Thus, we set

$$u_{R_2|S_1}(a|l) = \delta \quad \text{and} \quad u_{R_2|S_1}(r|l) = 1 - \delta.$$

Figure 7.2 illustrates the utility network corresponding to this problem.

The satisficing group welfare function represents the joint evaluation of all options available to the two decision makers as considered simultaneously from both their selectability and rejectability points of view and is given by

$$v_{S_1 S_2 R_1 R_2}(s_1, s_2, r_1, r_2) = u_{S_2|S_1}(s_2|s_1) u_{R_1}(r_1) u_{R_2|S_1}(r_2|s_1) u_{S_1}(s_1).$$

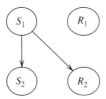

Figure 7.2. The utility network for the satisficing Ultimatum game.

7.2.1 Individually satisficing solutions

Once the satisficing group welfare function is defined, we may extract the individual selectability and rejectability functions for the responder by summing out all of the terms except the one pertaining to the self of interest, yielding

$$v_{S_2}(s_2) = \sum_{s_1} \sum_{r_1} \sum_{r_2} u_{S_2|S_1}(s_2|s_1) u_{R_1}(r_1) u_{R_2|S_1}(r_2|s_1) u_{S_1}(s_1)$$

$$= \begin{cases} 1 - \tau\delta & \text{for } s_2 = a \\ \tau\delta & \text{for } s_2 = r \end{cases} \qquad (7.4)$$

and

$$v_{R_2}(r_2) = \sum_{s_1} \sum_{s_2} \sum_{r_1} u_{S_2|S_1}(s_2|s_1) u_{R_1}(r_1) u_{R_2|S_1}(r_2|s_1) u_{S_1}(s_1)$$

$$= \begin{cases} \tau\delta & \text{for } r_2 = a \\ 1 - \tau\delta & \text{for } r_2 = r \end{cases} . \qquad (7.5)$$

Now recall that the satisficing set for a decision maker is the set of actions for which the selectability is at least as great as the product of q, the index of boldness, and the rejectability. Thus, setting $q = 1$ and comparing (7.2) with (7.3) and (7.4) with (7.5), the satisficing sets for the proposer and responder are

$$\Sigma_1(\tau,\delta) = \{a_1 \in \mathcal{A}_1 : v_{S_1}(a_1) \geq v_{R_1}(a_1)\} = \begin{cases} \{h\} & \text{if } \tau < \frac{1}{2-\delta} \\ \{l\} & \text{if } \tau > \frac{1}{2-\delta} \\ \{h,l\} & \text{if } \tau = \frac{1}{2-\delta} \end{cases}$$

$$\Sigma_2(\tau,\delta) = \{a_2 \in \mathcal{A}_2 : v_{S_2}(a_2) \geq v_{R_2}(a_2)\} = \begin{cases} \{a\} & \text{if } \tau < \frac{1}{2\delta} \\ \{r\} & \text{if } \tau > \frac{1}{2\delta} \\ \{a,r\} & \text{if } \tau = \frac{1}{2\delta} \end{cases} .$$

Figure 7.3 displays a cross-plot of τ versus δ. Let us first consider the proposer. Values of τ and δ that lie above the curve labeled $\tau = \frac{1}{2-\delta}$ (regions *I*

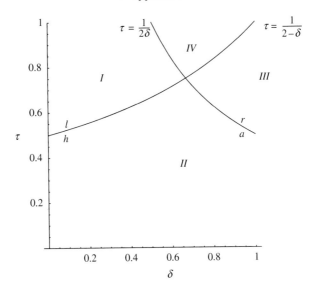

Figure 7.3. Cross-plot of intemperance (τ) versus indignation (δ) for individual decisions.

and IV) result in the proposer offering a low fraction, and values that lie below this curve result in a high fraction. For (τ, δ) pairs that lie on the line, both h and l are satisficing for the proposer. Next, consider the responder. Values of τ and δ that lie above the curve labeled $\tau = \frac{1}{2\delta}$ (regions III and IV) result in responder refusing the offer, and values that lie below this curve result in accepting the offer. For (τ, δ) pairs that lie on the line, both a and r are satisficing for the responder. The two curves divide the (τ, δ) square into four regions, corresponding to four different satisficing rectangles:

$$\mathcal{R}_{X_1 X_2}(\tau, \delta) = \Sigma_1(\tau, \delta) \times \Sigma_2(\tau, \delta) = \{(a_1, a_2): a_1 \in \Sigma_1(\tau, \delta) \text{ and } a_2 \in \Sigma_2(\tau, \delta)\}$$

$$= \begin{cases} \{(l,a)\} & \text{for } (\tau, \delta) \in I \\ \{(h,a)\} & \text{for } (\tau, \delta) \in II \\ \{(h,r)\} & \text{for } (\tau, \delta) \in III \\ \{(l,r)\} & \text{for } (\tau, \delta) \in IV \end{cases}.$$

In region I, a low fraction is offered and accepted, which is the Nash solution. It obtains when the proposer is prepared to risk failure and the responder is not easily offended. In region II, a high fraction is offered and accepted. This solution obtains because the proposer is either generous or fearful of rejection. In region III, a high fraction is offered and refused – a dysfunctional situation. In region IV, a low fraction is offered and rejected. Here, the proposer lets its greed get the better of its judgment, and the responder malevolently punishes it.

None of these results is particularly surprising, although it is interesting that the game, as parametrized, describes conditions where all four possible outcomes may occur. It is also interesting to note that the satisficing solution corresponds to the Nash solution only when the proposer is reckless (highly temerarious) and the responder is moderate (low indignation), which squares with intuition regarding human behavior.

7.2.2 Jointly satisficing solutions

In contrast to a competitive game (such as a zero-sum game) where one player's success is contingent on the other player's failure, the Ultimatum game is a game where either both players win (although not to the same degree) or both players lose (catastrophically, in this case). There is an intuitive operational group preference for both to win, but this "preference" is not explicitly coded into the utilities for either the conventional or the satisficing formulations of this game, nor is it evident in the way humans often play the game. Rather, it can only be an emergent phenomenon.

But if we do not impose an operational notion of preference on the group similar to the way we impose individual preferences (either unconditional or conditional), is there any hope that a group preference can be defined in such a way that individual actions, viewed from a group-level perspective, can be interpreted as a coherent group action? One possibility might be to view $\mathcal{R}_{X_1 X_2}$, the satisficing rectangle, as an aggregation of individual interests that represents group interest. Such an aggregation, however, has no stronger claim as a way to define group interests than does an aggregation of individual utilities to serve as a definition of group utility in the standard formulation of a game (an approach that is generally eschewed by game theorists). Fortunately, conditional satisficing game theory provides a purely mathematical means for group preferences to coexist with individual preferences in a coherent way.

To develop a mathematical notion of group preference, we first form the satisficing group welfare function, then extract the joint selectability and rejectability functions, and form the jointly satisficing set, yielding

$$\Sigma_{X_1 X_2}(\tau,\delta) = \{(a_1,a_2) \in \mathcal{A}_1 \times \mathcal{A}_2 \colon v_{S_1 S_2}(a_1,a_2) \geq q_{X_1 X_2} v_{R_1 R_2}(a_1,a_2)\},$$

where the joint selectability and joint rejectability functions are computed as

$$v_{S_1 S_2}(s_1,s_2) = \sum_{r_1}\sum_{r_2} u_{S_2|S_1}(s_2|s_1) u_{R_1}(r_1) u_{R_2|S_1}(r_2|s_1) u_{S_1}(s_1)$$

$$v_{R_1 R_2}(r_1,r_2) = \sum_{s_1}\sum_{s_2} u_{S_2|S_1}(s_2|s_1) u_{R_1}(r_1) u_{R_2|S_1}(r_2|s_1) u_{S_1}(s_1).$$

The resulting joint selectability and joint rejectability functions are

$$v_{S_1 S_2}(h,a) = 1 - \tau$$

$$v_{S_1 S_2}(h,r) = 0$$

$$v_{S_1 S_2}(l,a) = \tau - \tau \delta$$

$$v_{S_1 S_2}(l,r) = \tau \delta$$

and

$$v_{R_1 R_2}(h,a) = \tau^2 \delta - \tau^2 \delta^2$$

$$v_{R_1 R_2}(h,r) = \tau - \tau \delta - \tau^2 \delta + \tau^2 \delta^2$$

$$v_{R_1 R_2}(l,a) = \tau \delta - \tau^2 \delta + \tau^2 \delta^2$$

$$v_{R_1 R_2}(l,r) = 1 - \tau + \tau^2 \delta - \tau^2 \delta^2 .$$

Setting $q_{X_1 X_2} = 1$, the satisficing set for the group is obtained, as a function of (τ, δ), by comparing these functions for each joint action. Thus, the jointly satisficing set is a function of (τ, δ). Figure 7.4 partitions the (τ, δ) space into disjoint regions, with the corresponding jointly satisficing sets given by

$$\Sigma_{X_1 X_2}(\tau, \delta) = \begin{cases} \{(l,a)\} & \text{for } (\tau,\delta) \in A \\ \{(l,a),(l,r)\} & \text{for } (\tau,\delta) \in B \\ \{(l,r)\} & \text{for } (\tau,\delta) \in C \\ \{(l,a),(l,r),(h,a)\} & \text{for } (\tau,\delta) \in D \\ \{(l,r),(h,a)\} & \text{for } (\tau,\delta) \in E \\ \{(h,a),(l,a)\} & \text{for } (\tau,\delta) \in F \\ \{(h,a)\} & \text{for } (\tau,\delta) \in G \end{cases} \quad (7.6)$$

Notice that regions B, D, E, and F do not have unique solutions. Considering region B, for example, both joint options (l,a) and (l,r) are jointly satisficing for an (τ, δ) pair that lies in that region (highly intemperate, low to moderate indignation). The group, if it were to act as a single entity, would not reject either of these joint options – either would be good enough.

The jointly satisficing sets displayed in (7.6) indicate the degree of concordance among the players. We may also attach operational meanings to these sets as follows: For regions A, F, and G, we may deduce a group preference to consume the fortune. For region E, there is a group preference to be fair. For region D a possible group preference is to at least do something that is socially logical (such as the notion that punishment without cause is not rational behavior), as opposed to region C, where the only possible group preference is to be dysfunctional. Region B does not appear to offer a compelling operational notion of preference for the group.

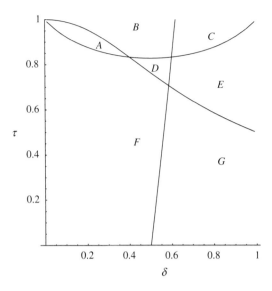

Figure 7.4. Cross-plot of intemperance (τ) versus indignation (δ) for group decisions.

7.2.3 Reconciling joint and individual choices

Once both individually and jointly satisficing decisions are defined, it remains to determine how (if it is even possible) they may be reconciled to each other to define a course of action that can be simultaneously viewed as acceptable to the group and to each of its members. Fortunately, this step is surprisingly straightforward in concept: We simply form the intersections of the satisficing rectangle and the jointly satisficing set to form the *compromise set*

$$\mathcal{C}_{X_1 X_2}(\tau,\delta) = \mathcal{R}_{X_1 X_2}(\tau,\delta) \cap \Sigma_{X_1 X_2}(\tau,\delta).$$

The compromise set comprises the joint actions that are simultaneously satisficing from both group and individual perspectives. If $\mathcal{C}_{X_1 X_2}(\tau,\delta) = \varnothing$, then there is no way to reconcile group and individual interests. Figure 7.5 superimposes the four regions corresponding to the satisficing rectangles and the seven regions corresponding to the jointly satisficing sets. By inspection, we compute the compromise set as

$$\mathcal{C}_{X_1 X_2}(\tau,\delta) = \begin{cases} (l,a) & \text{for } (\tau,\delta) \in I \cap (A \cup B \cup D \cup F) \\ (h,a) & \text{for } (\tau,\delta) \in II \cap (E \cup F \cup G) \\ \varnothing & \text{for } (\tau,\delta) \in III \\ (l,r) & \text{for } (\tau,\delta) \in IV \cap (B \cup C \cup E) \end{cases}.$$

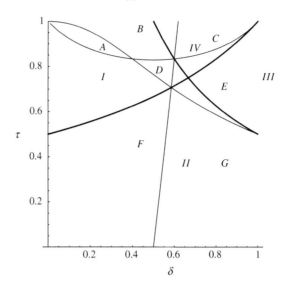

Figure 7.5. Comparison of group and individually satisficing decisions.

When computing the compromise set, we have relied only on the mathematical structure of the game and have not imposed any exogenous notions of group preference. Once the compromise set is defined, however, we are in a position to attempt to reconcile group and individual preferences in an operational sense. For all $(\tau, \delta) \notin III$, the compromise set is not empty; indeed, it contains exactly one element, indicating that there is a unique pair of individual choices that is coherent with the group choice. For region III, group and individual choices cannot be reconciled because the set of joint options that are satisficing for the group is empty.

7.3 The Stag Hunt game

The Stag Hunt game (also called the Trust Dilemma or the Assurance game) arises from a comment made by Rousseau in his *Discourse on Inequality*:

> If a group of hunters set out to take a stag, they are fully aware that they would all have to remain faithfully at their posts in order to succeed; but if a hare happens to pass near one of them, there can be no doubt that he pursued it without qualm, and that once he had caught his prey, he cared very little whether or not he had made his companions miss theirs. (Cited in (Ordeshook, 1986))

As usually formalized, the game involves two hunters. They can catch a stag only if they hunt the stag together, but each can catch a (much smaller) hare if

Table 7.3. *The payoff matrix for a*
two-player Stag Hunt game

	X_2	
X_1	s	h
s	(5, 5)	(1, 2)
h	(2, 1)	(2, 2)

they hunt separately. The players must individually decide between cooperation and noncooperation. This game therefore serves as a model for groups in which agents may benefit by cooperation, but for whom attempting to cooperate entails a significant risk of failure.

We first review the conventional two-player approach to the Stag Hunt game. Let s and h denote the decisions to hunt stag and hare, respectively. We assume that the payoff for hunting hare is independent of the other player; that is, both players must hunt stag to get the higher payoff, but each player can independently derive positive utility from hunting hare. The payoff matrix for this game is shown in Table 7.3.

There are two pure-strategy Nash equilibria for this game: (s,s) and (h,h). It is clear that (s,s) maximizes both players' payoffs, and thus is the desirable equilibrium point in the Pareto sense. However, a risk-averse player may still hunt hare, even though the payoff for hunting stag is greater. If, from the (s,s) equilibrium, one player deviates from stag hunting, the other player obtains the minimum payoff. Since one player's choice might impact the other player's payoff significantly and adversely, a conservative, yet still fully rational, choice for a risk-averse player would be to hunt hare.

Let us now reformulate this decision problem as a conditional satisficing game. For this development, we shall assume that utilities are completely dissociated. The four selves associated with this game are denoted S_i and R_i, with action spaces $\mathcal{A}_i = \{s,h\}$, $i = 1,2$. When formulating a satisficing game, our first task is to develop operational definitions for the selecting and rejecting preferences. We may define these preferences in many ways as long as the selectability is not merely a restatement of rejectability. Here, we choose to associate rejectability with the opportunity cost of an action. The opportunity cost of hunting hare is the payoff derived from catching stag, and the opportunity cost of hunting stag is the payoff for catching hare. Since stag hunting yields a greater resource, the opportunity cost for hunting hare is greater than the opportunity cost for hunting stag. We associate selectability with successful

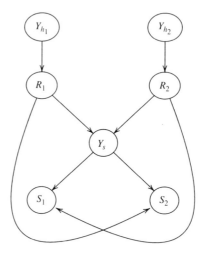

Figure 7.6. The utility/probability network for the Stag Hunt game.

cooperation; that is, if a successful stag hunt is available to the selecting self, it will prefer stag hunting, but if a successful stag hunt is not available, it will prefer hare hunting.

Given the operating definitions of selectability and rejectability, the next step is to define the interconnections between the four selves and form the utility network. The proposed model is illustrated in Figure 7.6. In addition to the deterministic vertices corresponding to the four selves, this network contains three stochastic vertices that are included to account for uncertainty. In general, it will not be certain that an attempted hare hunt will succeed, nor is it certain, even if both players hunt stag, that they will succeed. To account for the possibility of failure, we introduce three binary random variables: Y_s, Y_{h_1}, and Y_{h_2}. $Y_{h_i} = 1$ means that a successful hare hunt is available to the players *if they choose to pursue.* That is, a player who chooses to hunt hare will succeed if $Y_{h_i} = 1$ and fail if $Y_{h_i} = 0$. Similarly, $Y_s = 1$ signifies that a successful stag hunt is available to the selecting selves S_i, and $Y_s = 0$ means that stag hunting will result in failure.

Once the mass functions associated with each vertex and edge of the DAG are defined, the utility-probability network is complete. The satisficing group welfare function is

$$v_{S_1 S_2 R_1 R_2 Y_{h_1} Y_{h_2} Y_s}(s_1, s_2, r_1, r_2, b_{h_1}, b_{h_2}, b_s) = u_{S_1 | R_2 Y_s}(s_1 | r_2, b_s)$$

$$u_{S_2 | R_1 Y_s}(s_2 | r_1, b_s) p_{Y_s | R_1 R_2}(b_s | r_1, r_2) u_{R_1 | Y_{h_1}}(r_1 | b_{h_1})$$

$$u_{R_2 | Y_{h_2}}(r_2 | b_{h_2}) p_{Y_{h_1}}(b_{h_1}) p_{Y_{h_2}}(b_{h_2}).$$

Let ϕ_{s_i} and ϕ_{h_i} denote the utility of stag and hare, respectively, expressed in arbitrary units. Normalizing, the utility of hare hunting becomes

$$\mu_i = \frac{\phi_{h_i}}{\phi_{h_i} + \phi_{s_i}} = \frac{1}{1 + \frac{\phi_{s_i}}{\phi_{h_i}}}$$

for $i = 1, 2$, and we see that this utility depends only on the ratio $\frac{\phi_{s_i}}{\phi_{h_i}}$. The utility of staghunting is then $1 - \mu_i$.

Ostensibly, we might expect that $\phi_{s_i} = 5$ and $\phi_{h_i} = 2$, the payoff values given in Table 7.3, resulting in $\mu_i = \frac{2}{7}$. However, these values do not take into account the *attitudes* of the players. One of the core difficulties of the Stag Hunt is determining what players of differing risk-aversion levels should do. We may therefore introduce parameters, ρ_i, that expresses the degree of X_i's risk aversion. A player with $\rho_i = 1$ is risk-neutral, a player with $\rho_i > 1$ is risk-averse, and a player with $\rho_i < 1$ is payoff-seeking and tends to ignore risk. We then redefine μ_i as

$$\mu_i = \rho_i \frac{1}{1 + \frac{\phi_{s_i}}{\phi_{h_i}}}.$$

Thus, μ_i reflects a player's willingness to take risks as well as the relative utility for stag and hare. A maximally risk-averse player will hunt stag only if its success is certain, whereas a fully payoff-seeking player will hunt stag regardless of the odds. To ensure a meaningful game, we still require that both players will never prefer hare to stag, or $\mu_i < \frac{1}{2}$ for $i = 1, 2$.

These utilities assume that it is possible for X_i to hunt hare successfully. If, however, X_i cannot hunt hare successfully, there is no opportunity cost for stag hunting. We therefore define the conditional rejectability utilities as

$$u_{R_i|Y_{h_i}}(r_i|1) = \begin{cases} \mu_i, & \text{for } r_i = s \\ 1 - \mu_i, & \text{for } r_i = h \end{cases}$$

and

$$u_{R_i|Y_{h_i}}(u_i|0) = \begin{cases} 0, & \text{for } u_i = s \\ 1, & \text{for } u_i = h \end{cases}$$

for $i = 1, 2$. These conditional rejectabilities characterize the utility structures for hunting hare given that a hunt could or could not succeed. To compute the unconditional rejectability, we must take into consideration the probabilities of successful hunts and compute the expected utility of a successful hunt. These probabilities are characterized by the probability mass functions for Y_{h_i}:

$$p_{Y_{h_i}}(b_{h_i}) = \begin{cases} \eta_i, & \text{for } b_{h_i} = 1 \\ 1 - \eta_i, & \text{for } b_{h_i} = 0 \end{cases},$$

using the symbols Y for the random variable and b for its realization. The expected utility is obtained by summing over b_{h_i}, yielding

$$u_{R_i}(r_i) = \sum_{b_{h_i}} u_{R_i|Y_{h_i}}(r_i|b_{h_i}) p_{Y_{h_i}}(b_{h_i})$$

$$= \begin{cases} \epsilon_i, & \text{for } r_i = s \\ 1 - \epsilon_i, & \text{for } r_i = h \end{cases},$$

where $\epsilon_i = \mu_i \eta_i$ is the expected utility of hunting hare.

We next define the dependence relationships for Y_s. Here, the existence of separate, distinct selves S_i and R_i is critical. The distribution of this random variable, which is conditioned upon the rejecting preferences of both players, represents the degree to which a successful stag hunt is available to the selecting selves. Recall that the distribution is dependent on the preferences of only the rejecting selves, *not* what action the players implement, which is determined by the preferences of *both* selves. The distribution of Y_s incorporates both the degree to which R_1 and R_2 reject cooperation and how well the players hunt stag. We model the latter consideration by defining $0 \le \sigma \le 1$, which represents the probability of catching a stag given that the players cooperate. If neither R_1 or R_2 rejects stag hunting, then a successful stag hunt is possible – from the perspective of the selecting selves – with probability σ. We characterize this by defining

$$p_{Y_s|R_1 R_2}(b_s|h,h) = \begin{cases} \sigma, & \text{for } b_s = 1 \\ 1 - \sigma, & \text{for } b_s = 0 \end{cases}.$$

If, however, either player unilaterally rejects stag hunting, the probability of catching a stag is zero, thus

$$p_{Y_s|R_1 R_2}(b_s|s,s) = p_{Y_s|R_1 R_2}(b_s|s,h) = p_{Y_s|R_1 R_2}(b_s|h,s) = \begin{cases} 0, & \text{for } b_s = 1 \\ 1, & \text{for } b_s = 0 \end{cases}.$$

Notice that here we see praxeic preferences influencing the probability of an epistemic state, as discussed earlier. Since the player's rejecting preferences affect their willingness to hunt stag, this dependence structure is justifiable.

We compute the unconditioned mass function by summing over the conditional random variables, yielding

$$p_{Y_s}(b_s) = \sum_{r_1} \sum_{r_2} p_{Y_s|R_1,R_2}(Y_s|r_1,r_2) u_{R_1}(r_1) u_{R_2}(r_2)$$

$$= \begin{cases} \sigma(1-\epsilon_1)(1-\epsilon_2), & \text{for } b_s = 1 \\ 1 - \sigma(1-\epsilon_1)(1-\epsilon_2), & \text{for } b_s = 0 \end{cases}.$$

From p_{Y_s}, we see that as the expected hare-hunting utility decreases relative to the expected stag-hunting utility, the probability of a successful stag hunt increases. If both players completely prefer stag to hare ($\epsilon_1 = \epsilon_2 = 0$), the probability of the availability of a successful stag hunt is σ.

The selectability of each player is influenced by the probability of a successful stag hunt and the rejectability of the other player. This latter component permits each player to take into account the preferences of the other player when forming its preferences – an opportunity for altruism. To introduce altruism, we must specify circumstances in which the selecting self can accommodate the preferences of the other player's rejecting self; however, the structure of the Stag Hunt imposes limitations on how this may occur. If, for example, R_i fully rejects stag hunting, then S_j ($j \neq i$) can neither help nor hinder X_i's choice. Furthermore, if both R_1 and R_2 reject hare hunting, the probability for a successful stag hunt is high, and even egoistic selecting selves will select stag hunting.

A truly altruistic possibility occurs only when one rejecting self rejects stag hunting while the other rejecting self rejects hare hunting. If R_i rejects stag hunting and R_j does not, S_i may accommodate the other player by adjusting its preferences as influenced by R_j's preferences. Let $\delta_i \in [0,1]$ be X_i's *deference parameter*, such that $\delta_i = 0$ means that X_i is completely egoistic in its preferences. If $\delta_i > 0$, then X_i is *conditionally* altruistic, in the sense that it will suppress its egoistic preferences to the degree δ_i if, but only if, X_j rejects hare hunting. Otherwise, X_i will revert to its egoistic preferences. The corresponding conditional selectability model is

$$
u_{S_i | R_j Y_s}(s_i | r_i, b_s) =
\begin{cases}
1 & \text{for } u_i = s | r_j = s, b_s = 1 \\
0 & \text{for } u_i = h | r_j = s, b_s = 1 \\
1 & \text{for } u_i = s | r_j = h, b_s = 1 \\
0 & \text{for } u_i = h | r_j = h, b_s = 1 \\
0 & \text{for } u_i = s | r_j = s, b_s = 0 \\
1 & \text{for } u_i = h | r_j = s, b_s = 0 \\
\delta_i \sigma & \text{for } s_i = s | v_j = h, b_s = 0 \\
1 - \delta_i \sigma & \text{for } s_i = h | v_j = h, b_s = 0
\end{cases}
$$

for $i, j \in \{1, 2\}$, $i \neq j$. As a potentially altruistic scenario, assume that R_i rejects stag hunting but R_j does not. As long as σ is high, an altruistic X_i realizes that $b_s = 0$ because of its rejecting preferences. Its selecting self, S_i, can therefore accommodate the rejecting preferences of R_j by ascribing nonzero selectability to stag hunting.

The resulting marginal selectabilities are

$$u_{S_i}(a_i) = \begin{cases} \sigma(1-\epsilon_i)(1-\epsilon_j) + \delta_i\sigma(1-\epsilon_j)[1-\sigma(1-\epsilon_i)(1-\epsilon_j)] \\ \quad \text{for } a_i = s \\ 1 - \sigma(1-\epsilon_i)(1-\epsilon_j) - \delta_i\sigma(1-\epsilon_j)[1-\sigma(1-\epsilon_i)(1-\epsilon_j)] \\ \quad \text{for } a_i = h \end{cases}.$$

The satisficing rectangle is the set of action vectors that are simultaneously satisficing to each player individually, that is, for which $u_{S_i} \geq q u_{R_i}$, $i = 1, 2$. In Figure 7.7, we set q to unity and plot the regions of the satisficing rectangle as functions of ϵ_1 and ϵ_2. To account for all possible values of these parameters, we impose the maximum range $0 \leq \mu_i < \frac{1}{2}$, $i = 1, 2$. The expected utility must therefore obey this same constraint. Thus, only the region for which $0 \leq \epsilon_i < \frac{1}{2}$ is meaningful for the Stag Hunt. There are four possibilities for the satisficing rectangle. In Region (s, s), both players have low risk-aversion. In Region (h, h), both players consider the risk unacceptable and separately hunt hare. In Regions (h, s) and (s, h), however, one player is strongly risk-averse while the other is strongly payoff-seeking, with the result that one tries to cooperate and the other does not.

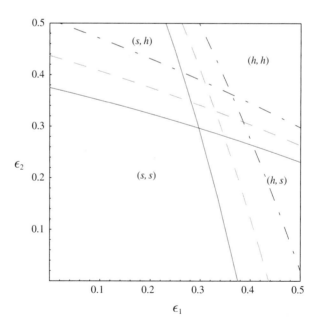

Figure 7.7. The satisficing rectangle sets for $\sigma = 0.6$, and $\delta_1 = \delta_2 = 0.0$ (solid line), $\delta_1 = \delta_2 = 0.25$ (dashed line), and $\delta_1 = \delta_2 = 0.5$ (dot-dashed line).

To understand these regions intuitively, consider the satisficing group under variations in ϵ_i. In Region (s,s), both players have determined that the relative benefits of catching a stag outweigh the risks inherent in cooperation. Next, consider Regions (h,s) and (s,h), where one of the players chooses to hunt hare while the other player, who is aware of the first player's reduced preference toward stag, nevertheless stands by its post. We might interpret this persistence as a manifestation of sufficient faith in the other player's eventual willingness to return to cooperation. Because players form their preferences on the preferences of others rather than their actions, this social network can endure moderate inconsistencies from its individual members. This is in stark contrast to the conventional Stag Hunt, in which similar shocks to individual utility invariably result in noncooperation. Of course, if either ϵ_1 or ϵ_2 increases too far, the benefits of stag hunting do not warrant further tolerance, and the society dissolves as both players hunt hare.

We also observe the significant impact of the deference parameter δ_i. As the players become increasingly deferential to each other's preferences, the cooperation region enlarges dramatically. In fact, for large δ_i's, it is possible for the cooperation region to exceed the boundaries for the Stag Hunt game (when $\epsilon_i > \frac{1}{2}$). That is, even if a player were actually to prefer hare to stag in terms of payoff, it would still defer to the other player if it strongly preferred stag.

7.4 The family walk

Suppose a family, consisting of a child (X_1), a mother (X_2), and a father (X_3), is to take one of three possible nature walks, denoted $\mathcal{A} = \{a, a', a''\}$. The child prefers an easy walk, the mother prefers beautiful scenery, and the father prefers long walks.

The first order of business in framing this scenario in the satisficing context is to settle on operational definitions for the notions of selectability and rejectability. From the point of view of each individual, the main goal is satisfaction according to its own criterion. Thus, it is reasonable to associate selectability with the degree of narrow self-interest. Accordingly, we define the three selectability functions in Table 7.4.

As the operational definition of rejectability, we assume that each agent has a unit of concern for the interests of others. Let us first consider the mother. Since she has concern for the interests of her child, she will encode this information in a rejectability function that is conditioned on the selectability conjecture of her child, as illustrated in Table 7.5. To interpret this table, consider the first column, which corresponds to $u_{R_2|S_1}(\cdot|a)$; that is, the child conjectures selecting a. Since this walk is tied for the most preferred outcome by the mother, she ascribes

Table 7.4. *Individual selectability functions*

Alternative	u_{S_1}	u_{S_2}	u_{S_3}
a	0.1	0.4	0.3
a'	0.3	0.4	0.6
a''	0.6	0.2	0.1

Table 7.5. *Mother's conditional rejectability* $u_{R_2|S_1}$

Alternative	$u_{R_2|S_1}(\cdot|a)$	$u_{R_2|S_1}(\cdot|a')$	$u_{R_2|S_1}(\cdot|a'')$
a	0.0	0.4	0.5
a'	0.4	0.0	0.5
a''	0.6	0.6	0.0

no conditional rejectability to that alternative and places all of her conditional rejectability mass on a' and a'' in inverse proportion to her selectability. Similar arguments apply if the child conjectures a' or a''.

The father's role in this decision process is first to defer to the conjectures of his child, then to the conjectures of his wife, and then, subject to those constraints, to reject the alternative that is least preferred in terms of his narrow self-interest. These values are provided in Table 7.6.

Finally, we must specify the child's rejectability. This rejectability is not conditioned, since the model does not call for the child's preferences to be influenced by the parents' preferences. Thus, the child's concern for the interests of others is neutral; that is, the child's rejectability function is uniform, given by

$$u_{R_1}(a) = u_{R_1}(a') = u_{R_1}(a'') = \frac{1}{3}.$$

Figure 7.8 illustrates the social influence flows of the satisficing praxeic network for the family walk.

To obtain a solution, we must compute the compromise set as defined in Section 6.6. Using the values provided in Tables 7.4, 7.5, and 7.6, and setting all q-values to unity, we may compute the social welfare function using (6.12),

Table 7.6. *Father's conditional rejectability* $u_{R_3|S_1S_2}$

	Alternative		
	a	a'	a''
$u_{R_3|S_1S_2}(\cdot|a,a)$	0	0	1
$u_{R_3|S_1S_2}(\cdot|a,a')$	0	0	1
$u_{R_3|S_1S_2}(\cdot|a,a'')$	0	1	0
	Alternative		
	a	a'	a''
$u_{R_3|S_1S_2}(\cdot|a',a)$	0	0	1
$u_{R_3|S_1S_2}(\cdot|a',a')$	0	0	1
$u_{R_3|S_1S_2}(\cdot|a',a'')$	1	0	0
	Alternative		
	a	a'	a''
$u_{R_3|S_1S_2}(\cdot|a'',a)$	0	1	0
$u_{R_3|S_1S_2}(\cdot|a'',a')$	1	0	0
$u_{R_3|S_1S_2}(\cdot|a'',a'')$	1	0	0

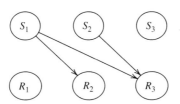

Figure 7.8. The utility network for the satisficing family walk problem.

yielding

$$w_{X_1X_2X_3}(a) = -0.05$$
$$w_{X_1X_2X_3}(a') = 0.36667$$
$$w_{X_1X_2X_3}(a'') = -0.052;$$

hence, the jointly satisficing set is $\Sigma_{X_1X_2X_3} = \{a'\}$.

We next compute the individually satisficing sets using (6.13), yielding $\Sigma_{q_1}^1 = \{a''\}$, $\Sigma_{q_2}^2 = \{a,a'\}$, and $\Sigma_{q_3}^3 = \{a'\}$. Consequently, $C_{X_1X_2X_3} = \varnothing$, and the society has not reached a compromise that is acceptable to all participants. However, if the child reduces q_1 to 0.9, then $\Sigma_{q_1}^1 = \{a', a''\}$, and a compromise exists with $C_{x_1 x_2 x_3} = \{a'\}$. An important feature of this example is that the child need only reduce its standards by a small amount to achieve a consensus. In terms of the narrow self-interest offering given by Table 7.4, we see, after taking into consideration the social dependencies that exist among the individuals, that the consensus alternative is best for the mother and the father and second best for the child.

7.5 Unmanned aerial vehicle control

Consider a group of three autonomous unmanned aerial vehicles (UAVs) ($n = 3$). The group is required to engage targets (either for reconnaissance or combat) while avoiding unnecessary exposure to hazards, collisions, and loss of communications. Each vehicle is able to detect all targets and hazards within its field of view, and each may communicate with its nearest neighbors, provided they are within range. Each UAV acts autonomously, but all are jointly responsible to achieve high performance and satisfy the constraints.

7.5.1 Simulation model

To simplify this coordination problem to its bare essentials, we assume the following:

C-1 The field of action consists of a grid divided into cells such that each target and each hazard is contained in one and only one cell. No cell may contain both a target and a hazard.

C-2 The vehicles fly at constant forward velocity but variable lateral velocity in a three-abreast formation. The forward velocity is $+1$ cell per time unit. The lateral velocity is drawn from the set $\mathcal{A} = \{-1, 0, 1\}$ cells per time unit, where negative signifies a move ahead and to the left; zero, a move straight ahead; and positive, a move ahead and to the right. Each cell may be occupied by at most one vehicle.

C-3 Each vehicle is able to detect all targets and hazards within a distance of d cells in the forward direction from their current cell, with unlimited lateral detection.

C-4 If a vehicle enters a cell that contains a target, the group scores one point.

C-5 If a vehicle enters a cell that contains a hazard, the group loses one point.

C-6 Direct communication may occur between two adjacent vehicles only if they are within D lateral cells of each other.

C-7 Indirect, or relayed, communication may occur between two nonadjacent vehicles if and only if a third vehicle can communicate directly with each of them.

C-8 Communicating vehicles may provide each other (a) their positions, (b) their current and anticipated action commands, and (c) negotiation proposals with respect to current and anticipated commands.

C-9 If vehicles cannot communicate, either directly or indirectly, they cannot know the positions or the actions (either instantiated or anticipated) of each other.

C-10 Vehicles may not collide; that is, they may not approach closer to each other than adjacent cells.

To evaluate our methodology, we adopt as a baseline the jointly optimal solution. We then consider two satisficing designs. The first provides a globally satisficing solution somewhat analogous to a globally optimal solution. The second approach illustrates how the satisficing methodology can be adapted to situations involving heterogeneous agents with partial knowledge or hierarchical structures. Together, the two satisficing solutions illustrate the flexibility of the satisficing design space and provide some idea of the trade-offs between structure, complexity, and performance. Note that all solutions assume a limited horizon, consistent with our earlier assumption of a limited forward field of view. Let us designate and order the positions of the the vehicles as X_1 for the leftmost vehicle, X_2 for the center vehicle, and X_3 for the rightmost vehicle.

7.5.2 Optimal solution

To obtain the optimal solution to the limited-horizon decision problem, we must compute the total target-minus-hazard count for each of the possible paths available to each agent, subject to the constraints. For a horizon of depth d, there is a total of 3^d paths available for each agent. This defines the *reachable cone*. Thus, if each of the three agents were to work strictly in isolation, we would require a total of 3^{d+1} calculations. However, to satisfy the constraints of avoiding collisions and loss of communication, we must calculate the costs for all joint paths between agents X_1 and X_2 and between X_2 and X_3. Note that, since X_1 and X_3 cannot violate constraints without consideration of X_2, it is not necessary to compute the costs of the joint paths between X_1, X_2, and X_3 simultaneously. The total number of arithmetic operations that must be performed to compute the jointly optimal solution is on the order of $2|\mathcal{A}|^{2d}$, where $|\mathcal{A}|$ is the cardinality of $\mathcal{A}$. For $|\mathcal{A}| = 3$ and $d = 3$, the number of operations is 1,458. Notice that this count is exponential in the horizon depth.

7.5.3 Satisficing solutions

The satisficing group welfare function for our hypothetical UAV system is a function of six variables, which we denote by $u_{S_1 S_2 S_3 R_1 R_2 R_3}(s_1, s_2, s_3, r_1, r_2, r_3)$, where s_i denotes an action available to S_i (i.e., the action that X_i might instantiate in the interest of its selectability) and r_i denotes an action available to R_i (i.e., the action that X_i might avoid instantiating in the interest of its rejectability). Our satisficing approach requires each agent to derive from the satisficing group welfare function a set of jointly satisficing decisions, that is, decisions that, from its perspective, would be satisficing for all agents. From this set of multipartite decision vectors, each player must determine which individual decision it should make. We assume that all agents are operating with the same model.

The satisficing group welfare function characterizes all of the relationships that exist between the six elements of the praxeic system. Once it is specified, it is straightforward to obtain the satisficing joint and individual solutions. Its specification, however, may be extremely complex. At each moment of (discrete) time, each vehicle has three possible actions at its disposal, and since each action must be considered in terms of both its selectability and its rejectability, the satisficing group welfare function has six arguments, each of which may assume any of three values, yielding a total of $3^6 = 729$ specifications. For many applications, however, we can reduce the number of independent parameters by exploiting social influence linkages between agents.

7.5.4 Global satisficing structure

For this case, we adopt the following operational definitions for the preferences for the actions available to each vehicle:

- The criterion for rejectability is to avoid hazards.
- The criterion for selectability is to seek targets while avoiding collisions or loss of communication.

For this problem, we assume that the joint rejectability criterion is independent of the joint selectability criterion. Thus, we may factor the satisficing group welfare function into the product of the joint selectability and joint rejectability functions as

$$v_{S_1 S_2 S_3 R_1 R_2 R_3}(s_1, s_2, s_3, r_1, r_2, r_3) = u_{S_1 S_2 S_3}(s_1, s_2, s_3) u_{R_1 R_2 R_3}(r_1, r_2, r_3). \quad (7.7)$$

If constraint violations are not imminent, then each agent operates with its own marginal selectability and rejectability functions, which are determined on the basis of the number of hazards and targets that can be encountered, respectively, for each action. If a constraint between two agents is imminent, then a joint selectability function is computed such that the constraints are honored between the two affected agents, while the third agent is free to operate without constraint. If a constraint violation involving all three agents is imminent, then a joint selectability function is computed for all three.

The computational burden for computing the marginal selectability and rejectability functions depends on the way the satisficing group welfare function is factored. If no factorization is employed, the total number of operations would be on the order of $|\mathcal{A}|^{2n} + (d+1)^2$. However, by factoring the satisficing group welfare function as indicated in (7.7), only $2|\mathcal{A}|^n + (d+1)^2$ operations are required per agent. Computing the jointly satisficing set requires an additional $|\mathcal{A}|^3$ operations per agent, with $|\mathcal{A}|$ operations per agent required to compute the individually satisficing set. Thus, the total number of operations per agent to compute the satisficing sets for $N = 3$, $|\mathcal{A}| = 3$ and $d = 3$ is $70 + 27 + 3 = 100$, which is much fewer than the number of operations required for the optimal solution. Notice that this complexity is quadratic, rather than exponential, in the horizon depth.

7.5.5 Markov satisficing structure

For this case, the definitions of the utilities are modified as follows:

- The rejectability of a move is determined by whether it leads to (a) a hazard, (b) a loss of communication, or (c) a collision.
- The selectability of a move is determined solely by whether it leads to a target.

More importantly, we wish to consider the effects of a Markovian constraint on agent interaction. In particular, we make the selectability and rejectability of a move contemplated by X_1 to be independent of the deliberations of X_3, conditioned on knowledge of X_2, and vice versa. For vehicle X_i, let us view the variable S_i as X_i viewed in terms of selectability, and R_i as X_i viewed in terms of rejectability.

There are many ways to approach the design of this system. As a general rule, it is reasonable to consider first the most critical attributes of the problem. We adopt here a conservative stance and assume that the rejectability attributes are more critical to the overall success of the mission than are the selectability attributes, since we assume that the number of hazards is greater than the number of targets.

Consider first rejectability. Clearly, each agent must place high rejectability on moves that would take it into cells occupied by hazards, and it must also avoid loss of communication and collisions. Because of the relative positioning of the vehicles, it is clear that X_2 cannot simultaneously avoid losing communication with both X_1 and X_3; similarly, it cannot simultaneously avoid collisions with these two agents. Thus, primary responsibility for these two rejectability criteria must fall to the wing agents. In other words, in addition to avoiding hazards, they must adjust their rejectability values to accommodate the actions of X_2. Thus, there is a natural flow of social influence from R_2 to both R_1 and R_3.

Let us now consider selectability. This criterion is concerned only with attacking targets. Typically, the opportunities to hit targets, especially with a sparse target scenario, will not be the same for all vehicles. As each agent looks ahead in its reachable cone over the next d moves, it can calculate the number of hits available to it. Let us designate the agent with the most number of possible hits as (for the current time only) the primary vehicle. The vehicle with the next most possible hits, given that the primary vehicle pursues its targets in an unconstrained manner, is the secondary vehicle. For the sake of illustration, let us assume that S_1 is primary, S_2 is secondary, and S_3 is tertiary. Accordingly, we will define a (temporally local) hierarchy that gives X_1 first priority in selecting its targets and gives X_2 second priority. Such a hierarchy is somewhat heuristic, since the sum of the hits possible for the other two vehicles, acting together unconstrained, may result in a total number of hits that exceeds what would otherwise be possible. Nevertheless, in the interest of simplicity, we adopt this as our design policy.

Our next step is to account for the interdependencies between rejectability and selectability. We will assume that rejectability and selectability *for a single agent* are independent; hence, there are no direct links between one agent's selectability self and its rejectability self. There may, however, be influence flows between selectability and rejectability between different agent selves. Let us consider how S_2 may be influenced by R_1 and R_3. If, for example, R_1 were to reject a move to the right, then a move to the right by S_2 could be detrimental to R_1 if it could result in a loss of communication. Consequently, it would be prudent for S_2 to conditionally defer to R_1 by being willing to reduce its selectability for moving to the right, even if hits are sacrificed. This is a manifestation of *conditional altruism* (a willingness to defer if so doing would actually benefit the other), as opposed to *categorical altruism* (unconditional abdication of one's individual interest in order to benefit another, regardless of whether the other would actually benefit). Thus, there is a natural flow of social influence from R_1 to S_2. By a similar argument, there is also a flow of social influence from R_3 to S_2. The overall flow of social influence is thus given by

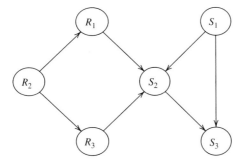

Figure 7.9. The utility network for the hierarchy $S_1 \to S_2 \to S_3$ UAV game.

Figure 7.9. The flows of influence for other hierarchies are defined similarly. A total of six different flow configurations would be possible. An important feature of this structure is that the system may be reconfigured dynamically as the relative positions of the agents in their environment evolves.

Figure 7.9 is a utility network corresponding to the factorization of the satisficing group welfare function as

$$v_{S_1 S_2 S_3 R_1 R_2 R_3}(s_1, s_2, s_3, r_1, r_2, r_3) = u_{S_1}(s_1) u_{S_3 | S_1 S_2}(s_3 | s_1, s_2) u_{S_2 | S_1 R_1 R_3}(s_2 | s_1, r_1, r_3)$$
$$u_{R_3 | R_2}(r_3 | r_2) u_{R_1 | R_2}(r_1 | r_2) u_{R_2}(r_2).$$

The structure of these conditional interdependencies is a function of the relative positions of the agents. If neither communication loss nor collision is imminent, then the conditional rejectabilities degenerate to the marginal ones, which are determined as a function of the minimum number of hazards that an agent could encounter for each of the actions. If a particular action were to violate a constraint, then the conditional rejectability would be adjusted to reject that action under the appropriate circumstances.

Concerning selectability, if the reachable cones for each of the agents do not intersect, then the joint selectability is the product of the individual selectability marginals. If the cones do intersect, then priority is given first to the primary agent and then to the secondary agent, with selectability being proportional to the maximum number of targets that are reachable with each action.

Once the joint and marginal selectability and rejectability utilities are computed, the joint and marginal satisficing sets may be constructed and a compromise solution identified. If, for the initial value of the negotiation index q, the compromise set is empty, then q may be iteratively lowered until the compromise set is not empty. In this way, the individual decisions are compatible

with group interests, although at the possible expense of some individual satisfaction as the players compromise in order to reach a decision that satisfies
both group and individual interests. Notice that this negotiation process does
not require the recalculation of the selectability and rejectability utilities, hence
is computationally very inexpensive.

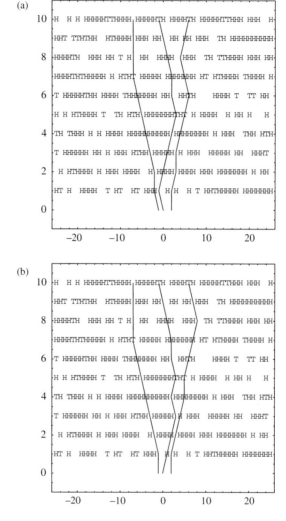

Figure 7.10. Simulated UAV trajectories: (a) satisficing design and (b) optimal
design.

Table 7.7. *Monte Carlo results: Global structure, Markov structure, and optimal solutions for horizon depth* $d = 3$

	Satisficing Global	Satisficing Markov	Optimal
Num targets	921	1026	950
Num hazards	654	882	686
Mean	2.67	1.44	2.64
Std dev	3.35	4.62	3.41

Table 7.8. *Monte Carlo results: Satisficing versus optimal, horizon depth* $= 2, 3, 4,$ *and* 5

Horizon depth	Sat mean	Sat std dev	Opt mean	Opt std dev
2	1.48	4.05	1.70	3.91
3	2.67	3.35	2.64	3.41
4	3.04	3.33	2.96	3.43
5	3.12	3.92	3.41	3.46

7.5.6 Simulation results

Figure 7.10 presents typical simulation results for a three-agent UAV system operating over ten time intervals and with a horizon depth of $d = 3$. The symbols "H" and "T" denote hazards and targets, respectively. Hazards and targets are generated randomly, with the probability of a hazard being 0.7 and the probability of a target being 0.1. The three line segments represent the trajectories of the three agents. Time flows from the bottom to the top of the figure along the ordinate, and the abscissa represents the lateral positions of the vehicles. Figure 7.10(a) illustrates the satisficing trajectory, and Figure 7.10(b) illustrates the optimal trajectory. Note that the hazard/target environment is the same. Table 7.7 displays the results from 100 Monte Carlo simulations for the optimal solution and both satisficing solutions. The initial conditions as well as hazard and target locations are generated randomly for each trial. The score for each trial is computed as the number of targets encountered minus the number of hazards encountered for the three agents. The sample means and standard deviations for the scores for both the satisficing and optimal solutions are indicated in the table.

Table 7.8 presents results for 100 Monte Carlo simulations for global structure and optimal solutions for horizon depths of 2, 3, 4, and 5. The difference between the satisficing and optimal averages is not statistically significant.

These results indicate (a) performance of the global structure satisficing solution is essentially the same as the performance of the optimal solution, and (b) the performance of the Markov satisficing solution is slightly worse than the performance of the global structure solution. This result is due to the tighter mode of coordination between the agents. The computational burden for both satisficing concepts is significantly less (approximately an order of magnitude for a depth of $d = 3$) than the computational burden for the optimal solution.

8

Conclusion

Knowledge is widely taken to be a matter of pedigree. To qualify as knowledge, beliefs must be both true and justified. Sometimes justification is alleged to require tracing of the biological, psychological, or social causes of belief to legitimating sources. Another view denies that causal antecedents are crucial. Beliefs become knowledge only if they can be derived from impecable first premises according to equally noble first principles. But whether pedigree is traced to origins or fundamental reasons, centuries of criticism suggest that our beliefs are born on the wrong side of the blanket. There are no immaculate preconceptions.

Where all origins are dark, preoccupation with pedigree is self-defeating. We ought to look forward rather than backward and avoid fixation on origins.

> — *Isaac Levi*
> *The Enterprise of Knowledge*
> *(MIT Press, 1980)*

Game theory has a great pedigree. To many, it is viewed as settled knowledge. In particular, the basic assumptions of categorical preference orderings and individual rationality have remained intact since the inception of the theory. Given these assumptions, the bulk of attention has focused on defining various solution concepts that conform to the rationality assumptions, including minimax theory, notions of equilibrium, coalition formation, principles of uncertainty, and analysis of repeated games. A fairly recent focus of interest has been advanced by the field of behavioral economics, which attempts to imbue games with greater psychological realism (for example, accounting notions of fairness and reciprocity). Yet another focus takes into account the computational limitations of human decision makers, leading to notions of bounded rationality. All of these threads, however, are ultimately connected to the fundamental mathematical structure of categorical utilities and the logical structure of individual rationality.

This book at least strains, if not breaks, those threads. In our treatment, categorical preference orderings are replaced by conditional preference orderings and individual rationality is replaced by a notion of simultaneous group and individual accommodation. Optimization is replaced by a softer, but still completely rigorous, notion of satisficing, and the door opens a bit wider to the design of principled negotiation strategies.

Tinkering with the fundamental structural assumptions of game theory is bound to be somewhat controversial. Indeed, it should be. Classical game theory enjoys a privileged place as a model of multistakeholder decision making. Its approach is simple, yet elegant, and surprisingly rich. However, it is constantly subjected to criticism. As acknowledged and then dismissed, Osborne and Rubinstein (1994) observe that "We do not discuss the assumptions that underlie the theory of a rational decision-maker. However, we do point out that these assumptions are under perpetual attack by experimental psychologists, who constantly point out severe limits to its application" (p. 5). Nevertheless, minimizing the concerns about the appropriate application of game theory is a luxury that perhaps can no longer be afforded.

While not challenging classical game theory when applied to situations for which the assumptions that underlie its structure are appropriate, we do challenge its uncritical application when those assumptions are in question. In doing so, we join with others who make a similar charge. Gintis (2009) argues that

> The most fundamental failure of game theory is its lack of a theory of when and how rational agents share mental constructs....Humans have a *social epistemology*, meaning that we have reasoning processes that afford us forms of knowledge and understanding, especially the understanding and sharing of the content our minds, that are unavailable to merely 'rational' creatures. This social epistemology characterizes our species. The bounds of reason are thus not the irrational, but the social [emphasis in original]. (p. xiv)

This book challenges results that have been decades in the making by researchers of the highest rank, and such a task should not be taken casually. But there is a compelling reason for doing so. Historically, most of the development of game theory has come fromt the social science disciplines and has focused on the design of mathematical tools to explain, predict, recommend, or justify human behavior. This book, however, arises from an engineering tradition, where the focus is on the synthesis of artificially intelligent decision-making entities. The difference in these two applications of the theory is profound. On the one hand, a mathematical model, no matter how sophisticated, can only approximate human behavior, and may or may not be accurate. On the other hand, the behavior of an artificially intelligent decision-making collective is *defined* by that mathematical description, and the collective is constrained to behave according to that model. Consequently, all of the desired social behavior must be explicitly included in the model and in the solution concepts.

As Kuhn (1962) has observed, however, an established theory is not likely to be displaced simply because it does not fully explain observed behavior. If it is to be challenged, the challenge must come from an alternative that provides a better fit to the facts. Conditional game theory presents that challenge. While certainly falling short of a comprehensive model of human behavior (nor is it intended to be), it provides an enlarged framework within which to develop more sophisticated social models than classical game theory permits that explicitly account for the social influence that agents (human or artificial) exert on each other.

Conditional game theory provides an opportunity to explore a number of theoretical issues, three of which are introduced in this book; namely, a theory of coordinatability, a joint utility-probability model to account simultaneously for preference and uncertainty, and a satisficing concept as an alternative to optimizing. However, as with any new theoretical development, there is much low-hanging fruit. Below are some samples of ideas that can be explored in the context of conditional game theory.

8.1 Learning

As discussed by Russell and Norvig (2003), a learning agent possesses structure that enables it to modify its behavior as it interacts with it environment. In addition to the performing element, the agent possesses three other components: a critic, a problem generator, and a learning element. The idea is for the critic to provide feedback regarding the quality of the performance, the role of the problem generator is to suggest alternatives, and the role of the learning element is to implement the changes to the system.

Let us consider our usual finite, strategic, noncooperative game setup, but now let us extend this structure to multiple stages defined as a time series of single-stage games, each with its own product action space (which may differ from stage to stage). As the players undertake their actions and assess their results, the opportunity to learn is presented. In a classical game-theoretic setup with categorical utilities, the only "knobs" an agent possesses are its definition of its utility and its solution concept. If the outcomes of the game at previous stages are not satisfactory to a player, it will be motivated to test other utility structures or solution concepts as suggested by its problem generator, even though they may lead to higher risk and lower payoffs in the short run. But in the long run, there is the possibility that the agent may discover a utility structure or a solution concept that improves its reward.

A well-known example of such a learning process occurs with the repeated-play Prisoner's Dilemma game. Learning players of that game would soon realize that reciprocity with punishment (the tit-for-tat strategy, for example) will generate a better outcome (mutual cooperation) than will playing the straight Nash equilibrium solution (mutual defection).

The learning possibilities with a conditional game, however, are greatly enlarged. In addition to the basic utility structure, the players can adjust their relationships with other agents by adding, deleting, or altering the influence linkages among the agents.

Another research possibility is to define a notion of group learning. Suppose a team of agents has a common goal and would benefit by coordination, but each must act autonomously. The coordination capacity provides a measure of the intrinsic ability of the group to work in harmony, regardless of the solution concepts its members employ. If the players could modify their conditional utilities in a way that increases their coordination capacity, and therefore their ecological suitability to function in their common environment, the possibility of improving performance should be enhanced. One way to approach this problem would be to appoint a group-level critic/problem generator (a coach) who would monitor the coordination capacity and suggest modifications to the players' influence relationships that would increase the capacity. For example, simulation studies could be performed to investigate the sensitivity of coordination capacity to perturbations in the parameters that define the utilities. To illustrate, consider the coordinated Prisoner's Dilemma game defined by Example 4.2. It is a conceptually simple matter to compute (either analytically or via finite differences) the partial derivative of $C(X_1, X_2)$ with respect to the model parameters α and β (see (4.4), (4.5), and (4.6)). The first-order sensitivity of the coordination capacity to, say, α, is then easily computed as

$$\Delta C = \frac{\partial C}{\partial \alpha} \Delta \alpha,$$

where ΔC and $\Delta \alpha$ denote changes in C and α, respectively. If $\Delta C > 0$, the coach would suggest that X_1 change its α parameter accordingly, thereby increasing the ability of the group to function harmoniously.

8.2 Model identification

The critical element of any decision-making enterprise is the model that characterizes the interests of the stakeholders and their relationships with each other. With classical game theory, these interests must all be encoded in the categorical utilities. Once defined, each agent applies its solution concept, thereby defining the outcome. When modeling decision making by human groups, it is easy to determine whether or not the observed behavior is consistent with the models predictions. From the standpoint of neoclassical economics, the baseline model is that each player is intent on maximizing its own benefit as defined by its payoff, regardless of the effect doing so has on others. When discrepancies arise, it is necessary to reexamine the model and solution concepts in an attempt to isolate and correct the discrepancy. A well-known example of such a practice is the introduction of parameters in a player's utility to account for such social issues as fairness or, as put by Fehr and Schmidt (1999), inequity aversion. If the modified model results in predicted behavior that agrees with observed behavior, we may conclude that an appropriate model has been identified.

When explicit influence relationships exist between stakeholders, the issue of model identification becomes more challenging since, not only is it necessary to determine appropriate utility values (essentially a parametric problem), but it is also necessary to determine appropriate influence linkages (a structure problem). This exercise is a reverse engineering problem to deduce appropriate relationships and to synthesize a mathematical model of the observed behavior that captures its essential features.

The field of system identification is an important control engineering subdiscipline that focuses on the synthesis of a mathematical model of a physical system that, given the system inputs, can reliably predict the system outputs. A great deal of statistical theory has been developed to address this problem. A potentially fruitful area of research would be to apply these methods, insofar as possible, to the problem of identifying a model that faithfully predicts observed behavior.

8.3 Dynamic systems

This book focuses exclusively on static games; that is, games that require a single action by each player. Although, in Section 8.1, we introduce the idea of a multistage game, the mathematical structure at each stage complies with the single-stage structure (repeated games, such as repeated-play Prisoner's

Dilemma, are a sequence of static games). It is not necessary, however, for a sequence of static games to have the same mathematical structure at each stage. For example, The UAV problem discussed in Section 7.5 is a sequence of static games, with each stage consisting of a single move by each player. At each stage, the environment changes, and the sets of actions and payoffs change accordingly. Although there may be an implicit relationship between the game definition at the tth stage and the $t + 1$st stage, the strategy at each stage is a single action by each player. With that structure, there is no explicit time-coupling between stages.

A dynamic game, by contrast, is a game that possesses an explicit time-dependency as the game is played. Such games often occur in engineering contexts. For example, consider the design of a distributed control system. A well-known design is the so-called model predictive control, a form of control in which the current control action is obtained by solving on-line, at each sampling instant, a finite horizon open-loop optimal control problem, using the current state of the plant as the initial state; the optimization yields an optimal control sequence, and the first control in this sequence is applied to the system.

Consider a collective of two coupled linear time-invariant systems with system equations

$$\mathbf{x}_1(t+1) = A_1\mathbf{x}_1(t) + B_{11}v_1(t)$$

$$\mathbf{x}_2(t+1) = A_2\mathbf{x}_2(t) + B_{21}v_1(t) + B_{22}v_2(t),$$

where A_i are 2×2 system matrices, B_{ij} are 2×1 input distribution matrices, $\mathbf{x}_i$ are two-dimensional state vectors, and v_i are scalar control functions, for $i, j \in \{1,2\}$. We assume that each X_i may choose its control function value from its action set $\mathcal{V}_i$, $i = 1, 2$.

The design of an appropriate control for the two subsystems may be viewed as a conditional game. The performance of each subsystem is determined by the value of the corresponding performance index, or utility. Clearly, X_1's performance index u_{X_1} is categorical, but X_2's performance index is conditional, and may be expressed in the form $u_{X_2|X_1}[\cdot|v_1(t)]$. The application of conditional game theory to this problem permits the definition of both group-level and individual-level performance measures, and conditional game-theoretic solution concepts can be applied to obtain a solution that is acceptable to the subsystems considered individually and to the collective considered jointly.

8.4 Final thoughts

In the preface I appealed to the metaphor, introduced by Novalis, that theories are nets. The motivation for applying that metaphor is that conditional game theory casts a wider net than does classical theory, thereby enabling us to characterize multistakeholder decision problems that involve explicit concepts of social influence.

It turns out that this new game-theoretic structure itself creates a net – actually, a network – and invites another analogy. Consider two models of fishing. With one, a number of fishermen independently cast their lines into the water, each individually hoping to catch as many fish as possible. Each fisherman's success however, depends not only on his own behavior but on the behavior (e.g., interference) of other fishermen. In other words, the fishermen are participants in a classical game. Each fisherman (player) applies his own fishing method (his action) designed to achieve a given success (his payoff), and the number of fish caught by each individual fisherman is dependent on the behavior of all fishermen (the outcome). Each fisherman, however, is intent on catching as many fish as possible, regardless of the success of other fisherman (he is motivated by self-interest). Graphically, this game can be viewed as a set of vertices, one for each fisherman, but with no edges connecting them – a degenerate network.

Now consider a second model, where the fishermen jointly cast a net with the goal of catching multiple fish together. This model is analogous to a group of socially interconnected decision makers. There is a social linkage (actually a physical linkage) between them as each controls his portion of the net according to his own desires, which are influenced by the desires, as well as the actions, of others. The structure of the net and the skill of the fishermen create a sophisticated society that possesses an intrinsic capability to coordinate which is completely lacking, at least ostensibly, with the first model. This group corresponds to a graph with edges – a network. And, unlike the first model, it naturally admits a notion of group preference and group rationality.

This book develops a mathematical structure to accommodate decision scenarios characterized by the second model. To do so, it focuses on the foundational issues of multistakeholder decision making. The key element of this approach, which runs through all of the development, is the application of the mathematical syntax of probability theory to nonprobabilistic issues.

As the influence of game theory becomes more pronounced in the engineering and computer science communities, the application areas will enlarge and the interest in synthesizing multiagent systems and distributed control systems will increase. In part, this book is an invitation to researchers, not only in those disciplines but in the social sciences as well, to view multistakeholder decision making through the more socially accommodating lens that conditional game theory affords.

Appendix A
Probability

One cannot escape the feeling that these mathematical formulas have an independent existence and an intelligence of their own, that they are wiser than we are, wiser even than their discoverers, that we get more out of them than was originally put into them.

— Heinrich Hertz

A.1 The uses of probability theory

Whenever one talks of foundational assumptions, it is hard to escape addressing philosophical issues. One area of mathematics that has long captured the interest of philosophers is probability theory. As betrayed by its very name, probability theory is typically applied to the epistemological issue of quantifying uncertainty regarding phenomena for which precise knowledge is not available.

In this book, we appropriate the mathematical structure and syntax of probability theory for a praxeological application. By so doing, we move far afield from its traditional epistemological home. Keeping in mind that this interpretation is nontraditional and may be controversial, we take considerable pains to provide a principle-based justification for appropriating probability theory for a nonepistemological application. As Hamming aptly observed, "it is dangerous to apply any part of science without understanding what is behind the theory" (Hamming, 1991, p. viii). Applying probability theory is essentially an art form and must be used with judgment and skill. The main difference between traditional usage and our usage is that, whereas probability is traditionally painted on an epistemological canvas, we choose to paint on a praxeological canvas as well.

The vast field of epistemology focuses on analyzing how knowledge relates to truth, belief, and justification. Probability theory is an important tool for the study of these concepts, since it provides a systematic mechanism with which to evaluate the evidence regarding the truth of a proposition, to combine different pieces of evidence, and to relate one piece of evidence to another piece of evidence.

The origins of probability theory, however, are somewhat more humble. In their book *The Empire of Chance*, Gigerenzer et al. (1989) provide a summary of the issues that motivated people to consider questions in the presence of incomplete knowledge. They conclude that the origins of probabilistic thinking are not only ancient but murky. However, beginning with Pascal and continuing with Daniel Bernoulli, as well as many others, a calculus of expectations developed based on notions of the chance of an outcome and of its value. Considering expectations provided a way to define a systematic framework within which to evaluate choices on the basis of the likelihood of and value of the consequences.

Probability theory has been well established as a useful mathematical tool with which to model uncertainty. In this role, it has been both lauded and criticized. It has been subjected to many different interpretations, with the best known being (a) frequency interpretations, (b) logical interpretations, (c) personal interpretations, (d) propensity interpretations, and (v) subjective interpretations. When studying a mathematical concept, however, it is always useful to isolate the mathematics from the application. Focusing on applications sometimes imputes a physical, logical, social, or personal significance to a mathematical object that may or may not apply. And even if it were to apply, it would have no bearing on the mathematics.

Probability theory was placed on a solid axiomatic footing by Kolmogorov (1933), who defined probability in terms of measure theory and established it as an ahistorical and acontextual mathematical study, free of all issues that motivated previous studies. He thereby rendered probability theory free from any interpretations, such as a way to account for uncertainty. As von Plato (1994) observed, "a mathematician would answer the question of what is probability by saying: Anything that satisfies the axioms" (p. 1).

Even more to the point, whether or not probability theory is used as a mathematical model to describe phenomena in a context entirely different from dealing with randomness depends on whether or not the mathematical structure of that model is appropriate for the application. However, the very name "probability theory" tends to restrict the application of the mathematics, although this name stems more from its historical beginnings than from its mathematical structure.

The power of probability theory is succinctly expressed by Glenn Shafer (cited in, Pearl, 1988, p. 15): "probability is not really about numbers; it is about the structure of reasoning." The critical structural components that make probability theory a useful model for reasoning are conditioning and independence (without those features, probability theory would be little more than a subtopic of measure theory). And these are exactly the features that are exploited in the development of conditional utilities.

Thus, it might behoove us to move beyond the narrow application of probability, as it is traditionally used, to consider an expanded view. To motivate such an endeavor, consider the following analogy: Newton's observations regarding the behavior of moving objects in a gravitational field motivated the development of differential equation theory, but to refer to that mathematical field as, say, *motion theory*, because of its historical origins, would be entirely too limiting. No one would suggest that physical mechanics "owns" differential equation theory. It is body of mathematics that stands on its own without reference to any particular application. Similarly, we should not suppose that the field of statistics, or any other intellectual discipline, "owns" probability theory. Statistics just happens to benefit from the application of that body of mathematics.

The fact that many consider probability theory to be a good model for the way people reason in the presence of uncertainty is similar to the fact that people consider differential equation theory a good model for the way moving objects behave in a gravitational field. Both are nothing more than mathematical constructs that conform to observed behavior. The fact that mankind has devised precise mathematical descriptions of observed behavior (either natural or human) is a remarkable accomplishment. As the physicist Richard Feynman has observed, "mathematics is just organized reasoning" (Feynman, 1967, p. 41). Both differential equation theory and probability theory are powerful systematic reasoning tools, but these tools do not force behavior. Newton's "laws" do not dictate the motion to an object, and probabilistic "laws" do not dictate the way people combine evidence to form opinions and beliefs, nor do these "laws" dictate the outcome of coin toss or any other random phenomenon.[1] Such models serve only to explain and predict behavior. Since they do so effectively, we may be tempted to view them as "laws" that somehow

---

[1] de Finetti (1990) began his book *Theory of Probability* with the rather provocative phrase "PROBABILITY DOES NOT EXIST." Essentially, he means that probability is a subjective way to model uncertainty. It is not an intrinsic property of an experiment. For example, although a coin may possess physical attributes of mass, dimension, and a moment of inertia, it *does not* possess a physical attribute of, say, $p = \frac{1}{2}$, meaning that the probability of landing heads when tossed is one-half. This assertion is a model of behavior that one may impose on the experiment of tossing a coin; it is not a property of the coin itself.

impose behavior on the phenomenon. But they are not laws. They are nothing more than models.

A.2 Probability distribution theory

In Chapter 2 we define a concordant utility and argue that it qualifies as a mass function that possesses all of the properties of a probability mass function, including marginalization, the chain rule, and Bayes rule. In order to justify this argument, however, we are under obligation to provide a rigorous justification from the perspective of probability theory.

The key issue under study is the fact that the concordant utility $U_{\boldsymbol{x}_k}$ (see Section 2.2.2 for definitions and notation), although it possesses the syntactical structure of a mass function, is not in the form of a standard multivariate probability mass function. The main difference is that, whereas a standard joint probability mass function for a set of k random variables is a function of k independent variables, the concordant utility for a set of k stakeholders is a function of kn independent variables.

Our challenge is to construct a probability model that corresponds to the structure of the concordant utility. We approach this problem in two steps. First, we construct a conventional probability model to serve as an analogy for a multivariate utility function (even though we do not depend on such entities in the theoretical development provided in Chapter 2). We then construct a probability model to serve as an analogy to a concordant utility.

A.2.1 Foundational elements

We begin by reviewing the basic foundational elements of probability theory. We first provide fundamental definitions and set the notation. A *sample space* Ω is a collection of outcomes of an experiment. A *sigma field* $\mathcal{F}$ is collection of subsets of Ω that contains the sample space and is closed under countable unions and complementation. A *probability measure* P is a nonnegative function mapping members of $\mathcal{F}$ to the unit interval such that $P(\Omega) = 1$ and, if $A_1, A_2, \ldots$ is a countable set of mutually exclusive members of $\mathcal{F}$, then $P(\cup_i A_i) = \sum_i P(A_i)$. The triple $(\Omega, \mathcal{F}, P)$ is called a *probability space*.

Consider a set of k sample spaces Ω_i, $i = 1, \ldots, k$, let $\mathcal{F}_i$ denote a sigma field defined over the subsets of Ω_i, and let P_i denote a probability measure over $\mathcal{F}_i$. We may form the *product sample space* as the Cartesian product of the individual sample spaces, denoted $\boldsymbol{\Omega} = \Omega_1 \times \cdots \times \Omega_k$, and the product sigma field, denoted $\mathcal{F}$, as the smallest sigma field that contains all *measurable rectangles* $A_1 \times \cdots \times A_k$, where $A_i \in \mathcal{F}_i$. Finally, let $\mathcal{P} \colon \mathcal{F} \to [0, 1]$ be a probability measure such that, for all $A_i \in \mathcal{F}_i$,

$$P_i(A_i) = \mathcal{P}(\Omega_1 \times \cdots \times \Omega_{i-1} \times A_i \times \Omega_{i+1} \times \cdots \times \Omega_k). \qquad (A.1)$$

The triple $(\mathbf{\Omega}, \mathcal{F}, \mathcal{P})$ is termed a *product probability space*, For a discussion of product probability spaces, see Neveu (1965).

Although probability spaces constitute the fundamental structural elements of a probability scenario, dealing directly with such spaces is cumbersome. A much more convenient quantity to work with is the distribution function of a random variable. A *random variable* Z is a Borel measurable function[2] $Z: \Omega \to \mathbb{R}$. The distribution function of Z, denoted $F_Z: \mathbb{R} \to [0,1]$, corresponds to the probability that Z takes values less than or equal to $z \in \mathbb{R}$; that is,

$$F_Z(z) = P[\{\omega \in \Omega: Z(\omega) \in (-\infty, z]\}].$$

A random variable is said to be discrete if it can assume only a countable number of values, in which case we may define a *probability mass function* p_Z as the probability that the random variable equals z; that is,

$$p_Z(z) = P[\{\omega \in \Omega: Z(\omega) = z\}].$$

A *random vector* $\mathbf{Z}$ is a collection of $n > 1$ random variables, all defined over the same probability space; that is, $\mathbf{Z}: \Omega \to \mathbb{R}^n$. Let $\mathbf{z} = (z_1, \ldots, z_n) \in \mathbb{R}^n$. The distribution function of $\mathbf{Z}$, denoted $F_Z: \Omega \to [0,1]$, corresponds to the probability that $\mathbf{Z}$ takes values in the *corner set*

$$C(\mathbf{z}) = (-\infty, z_1] \times \cdots \times (-\infty, z_n],$$

yielding the distribution function

$$F_{\mathbf{Z}_i}(\mathbf{z}_i) = P_i[\omega_i \in \Omega_i: \mathbf{Z}_i(\omega_i) \in C(\mathbf{z}_i)\}],$$

If $\mathbf{Z}$ is discrete, then it can assume only a countable set of values, and the corresponding probability mass function becomes

$$p_{\mathbf{Z}_i}(\mathbf{z}_i) = P_i[\omega_i \in \Omega_i: \mathbf{Z}_i(\omega_i) = \mathbf{z}_i\}].$$

A.2.2 Distributions for collections of random variables

Let $\{Z_1, \ldots, Z_k\}$ be a collection of k discrete random variables, respectively defined over the probability spaces $(\Omega_i, \mathcal{F}_i, P_i)$; that is, $Z_i: \Omega_i \to \mathbb{R}, i = 1, \ldots, k$.

The multivariate, or joint, distribution function for the collection $\{Z_1, \ldots, Z_k\}$ is a function, denoted $F_{Z_1 \cdots Z_k}: \mathbb{R}^k \to [0, 1]$, and corresponds to the probability

---

[2] A function is *measurable* if the inverse image of open sets in the range space is a member of the sigma field in the domain space.

that each Z_i takes values less than or equal to z_i, $i = 1, \ldots, k$, yielding

$$F_{Z_1 \cdots Z_k}(z_1, \ldots, z_k) = \mathcal{P}[\{(\omega_1, \ldots, \omega_k) \in \mathbf{\Omega} : Z_i(\omega_i) \in (-\infty, z_i], \ i = 1, \ldots, k\}].$$

If all of the Z_i's are discrete, then the corresponding probability mass function is

$$p_{Z_1 \cdots Z_k}(z_1, \ldots, z_k) = \mathcal{P}[\{(\omega_1, \ldots, \omega_k) \in \mathbf{\Omega} : Z_i(\omega_i), = z_i, i = 1, \ldots, k\}].$$

Since (A.1) is assumed to hold, the probability mass function for each Z_i is given by marginalization:

$$p_{Z_i}(z_i) = \sum_{\sim z_i} p_{Z_1 \cdots Z_k}(z_1, \ldots, z_k),$$

where the notation $\sum_{\sim z_i}$ means that the sum is taken over all arguments except z_i.

If the collective $\{Z_1, \ldots, Z_k\}$ is mutually independent, then

$$p_{Z_1 \cdots Z_k}(z_1, \ldots, z_k) = \prod_{i=1}^{k} p_{Z_i}(z_i).$$

Conditional probability mass functions are defined as

$$p_{Z_k | Z_j}(z_k | z_j) = \frac{p_{Z_k Z_j}(z_k, z_j)}{p_{Z_j}(z_j)}.$$

Finally, by Bayes rule,

$$p_{Z_j | Z_k}(z_j | z_k) = \frac{p_{Z_k | Z_j}(z_k | z_j) p_{Z_j}(z_j)}{p_{Z_k}(z_k)}.$$

A.2.3 Distributions for collections of random vectors

Let $\{\mathbf{Z}_1, \ldots, \mathbf{Z}_k\}$ be a k-member collection of n-dimensional discrete random vectors, each of which is defined over a distinct probability space; that is, $\mathbf{Z}_i : \Omega_i \to \mathbb{R}^n$. Let $\mathbf{z}_i = (z_{i1}, \ldots, z_{in})$. The joint distribution function of $\{\mathbf{Z}_1, \ldots, \mathbf{Z}_k\}$ is defined as

$$F_{\mathbf{Z}_1 \cdots \mathbf{Z}_k}(\mathbf{z}_1, \ldots, \mathbf{z}_k) = \mathcal{P}[\{(\omega_1, \ldots, \omega_k) \in \mathbf{\Omega} : \mathbf{Z}_i(\omega_i) \in C(\mathbf{z}_i), \ i = 1, \ldots, k\}],$$

and the corresponding joint probability mass function is

$$p_{\mathbf{Z}_1 \cdots \mathbf{Z}_k}(\mathbf{z}_1, \ldots, \mathbf{z}_k) = \mathcal{P}\{(\omega_1, \ldots, \omega_k) \in \mathbf{\Omega} : \mathbf{Z}_i(\omega_i) = \mathbf{z}_i, \ i = 1, \ldots, k\}.$$

Since, as before, (A.1) is assumed to hold, the probability mass function for each $\mathbf{Z}_i$ is given by marginalization:

$$p_{\mathbf{Z}_i}(\mathbf{z}_i) = \sum_{\sim \mathbf{z}_i} p_{\mathbf{Z}_1 \cdots \mathbf{Z}_k}(\mathbf{z}_1, \ldots, \mathbf{z}_k).$$

If the collective $\{\mathbf{Z}_1, \ldots, \mathbf{Z}_k\}$ is mutually independent, then

$$p_{\mathbf{Z}_1 \cdots \mathbf{Z}_k}(\mathbf{z}_1, \ldots, \mathbf{z}_k) = \prod_{i=1}^{k} p_{\mathbf{Z}_i}(\mathbf{z}_i).$$

Conditional probability mass functions are defined as

$$p_{\mathbf{Z}_k|\mathbf{Z}_j}(\mathbf{z}_k|\mathbf{z}_j) = \frac{p_{\mathbf{Z}_k \mathbf{Z}_j}(\mathbf{z}_j, \mathbf{z}_k)}{p_{\mathbf{Z}_j}(\mathbf{z}_j)}.$$

Finally, by Bayes rule,

$$p_{\mathbf{Z}_j|\mathbf{Z}_k}(\mathbf{z}_j|\mathbf{z}_k) = \frac{p_{\mathbf{Z}_k|\mathbf{Z}_j}(\mathbf{z}_k|\mathbf{z}_j) \, p_{\mathbf{Z}_j}(\mathbf{z}_j)}{p_{\mathbf{Z}_k}(\mathbf{z}_k)}.$$

An interesting feature of this structure is that the "marginal" distribution, although associated with a single random phenomenon (i.e., only one elementary event ω_i has been instantiated), it is still a function of k independent variables. Consequently, we can evaluate the marginal probability that the $i l$th component of $\mathbf{z}_i$ is instantiated, yielding

$$p_{z_{il}}(z_{il}) = \sum_{\sim z_{il}} p_{\mathbf{Z}_i}(\mathbf{z}_i).$$

We thus see that a concordant utility $U_{\mathcal{X}_k}(\mathbf{a}_1, \ldots, \mathbf{a}_k)$, where $\mathbf{a}_i = (a_{i1}, \ldots, a_{in})$, $i = 1, \ldots, k$, is mathematically equivalent to the joint distribution of a k-element family of n-dimensional random vectors. We thus establish that the syntactic structure of a concordant utility admits praxeological analogues to the traditional probabilistic concepts of conditioning, independence, marginalization, the chain rule, and Bayes rule.

Bibliography

A. E. Abbas. The algebra of utility inference. *CoRR*, cs.AI/0310044, 2003.

A. E. Abbas. From Bayes' nets to utility nets. *Proceedings of the 29th International Workshop on Bayesian Inference and Maximum Entropy Methods in Science and Engineering*, pages 3–12, 2009.

A. E. Abbas and R. A. Howard. Attribute dominance utility. *Decision Analysis*, 2(4): 185–206, 2005.

N. H. Abel. Untersuchung der functionen zweier unabhängig veränderlichen Gröszen x und y, wie $f(x,y)$ welche die Eigenschaft haben, dasz $f[z, f(x,y)]$ eine symmetrische function von z, x, und y ist. *J. reine u. angew. Math. (Crelle's Journal)*, 1:11–15, 1826.

J. Aczél. *Lectures on Functional Equations and Their Applications*. Academic Press, New York, 1966.

K. J. Arrow. *Social Choice and Individual Values*. John Wiley, New York, 1951. 2nd ed., 1963.

K. J. Arrow. *The Limits of Organization*. W. W. Norton & Company, New York, 1974.

K. J. Arrow. Rationality of self and others in an economic system. In R. M. Hogarth and M. W. Reder, editors, *Rational Choice*. University of Chicago Press, Chicago, 1986.

R. Axelrod. *The Evolution of Cooperation*. Basic Books, New York, 1984.

R. Axelrod and W. D. Hamilton. The evolution of cooperation. *Science*, 211:1390–1396, 1981.

K. Bailey. *Social Entropy Theory*. State University of New York Press, Albany, NY, 1990.

S. Benn. Rationality and political behaviour. In G. W. Mortimore and S. I. Benn, editors, *Rationality and the Social Sciences*. Routledge and Kegan Paul, London, 1976.

S. I. Benn. The problematic rationality of political participation. In P. Laslett and J. Fishkin, editors, *Philosophy, Politics and Society*. Blackwell, Oxford, 1979.

C. Bicchieri. *Rationality and Coordination*. Cambridge University Press, Cambridge, 1993.

K. G. Binmore. *Game Theory and the Social Contract: Just Playing*. MIT Press, Cambridge, MA, 1998.

G. E. Bolton and A. Ockenfels. A stress test of fairness measures in models of social utility. *Economic Theory*, 24(4), 2005, pages 957–982.

T. Burns and G. M. Stalker. *The Management of Innovation*. Tavistock Publications, London, 1961.

C. Camerer. *Behavioral Game Theory: Experiments in Strategic Interaction*. Princeton University Press, Princeton, NJ, 2003.

C. Camerer, G. Lowenstein, and M. Rabin, editors. *Advances in Behavorial Economics*. Princeton University Press, Princeton, NJ, 2004a.

C. Camerer et al. *Foundations of Human Sociality: Economic Experiments and Ethnographic Evidence from Fifteen Small-scale Societies*. Oxford University Press, Oxford, 2004b.

R. W. Cooper. *Coordination Games*. Cambridge University Press, Cambridge, 1999.

T. M. Cover and J. A. Thomas. *Elements of Information Theory*. John Wiley, New York, 1991.

R. G. Cowell, A. P. Dawid, S. L. Lauritzen, and D. J. Spiegelhalter. *Probabilistic Networks and Expert Systems*. Springer Verlag, New York, 1999.

R. T. Cox. Probability, frequency, and reasonable expectation. *American Journal of Physics*, 14:1–13, 1946.

F. G. Cozman. Credal networks. *Artificial Intelligence*, 120:199–233, 2000.

B. de Finetti. La prévision: ses lois logiques, ses sources subjectives. *Annales de l'Institut Henri Poincaré*, 7:1–68, 1937. Translated as Forsight. Its logical laws, its subjective sources, in H. E. Kyburg Jr. and H. E. Smokler, editors, *Studies in Subjective Probability*, Wiley, New York, 1964, pages 93–158.

B. de Finetti. *Theory of Probability*, volume 1. Wiley Interscience, New York, 1990.

J. Dewey. *Logic: The Theory of Inquiry*. Henry Holt and Company, New York, 1938.

F. Y. Edgeworth. *Mathematical Psychics*. C. Kegan Paul & Co., London, 1881.

H. J. Einhorn and R. M. Hogarth. Decision making under ambiguity. In R. M. Hogarth and M. W. Reder, editors, *Rational Choice*. University of Chicago Press, Chicago, 1986.

A. Einstein. On the method of theoretical physics. *Philosophy of Science*, 1(2):163–169, 1934.

J. Elster, editor. *The Multiple Self*. Cambridge University Press, Cambridge, 1985.

J. Elster, editor. *Rational Choice*. Basil Blackwell, Oxford, 1986.

E. Fehr and S. Gächter. Cooperation and punishment. *American Economic Review*, 90(4):980–994, 2002.

E. Fehr and K. Schmidt. A theory of fairness, competition, and cooperation. *Quarterly Journal of Economics*, 114:817–868, 1999.

R. Feynman. *The Character of Physical Law*. MIT Press, Cambridge, MA, 1967.

P. C. Fishburn. *The Theory of Social Choice*. Princeton University Press, Princeton, NJ, 1973.

H. G. Frankfurt. Freedom of the will and the concept of a person. *Journal of Philosophy*, 68:5–20, 1971.

M. Friedman. *Price Theory*. Aldine Press, Chicago, 1961.

J. Gale, K. Binmore, and L. Samuelson. Learning to be imperfect: The ultimatum game. *Games and Economic Behavior*, 8:56–90, 1995.

N. Georgescu-Roegen. *The Entropy Law and the Economic Process*. Harvard University Press, Cambridge, MA, 1971.

G. Gigerenzer, Z. Swijtink, T. Porter, L. Daston, J. Beatty, and L. Krüger. *The Empire of Chance*. Cambridge University Press, Cambridge, 1989.

H. Gintis. *The Bounds of Reason: Game Theory and the Unification of the Behavioral Sciences*. Princeton University Press, Princeton, NJ, 2009.

U. Gneezy. Deception: The role of consequences. *American Economic Review*, 95(1): 384–394, 2005.

R. E. Goodin. Cross-cutting cleavages and social conflict. *British Journal of Political Science*, 5:516–519, 1975.

R. E. Goodin. Laundering preferences. In J. Elster and A. Hylland, editors, *Foundations of Social Choice Theory*, chapter 3, pages 75–101. Cambridge University Press, Cambridge, 1986.

S. Goyal. *Connections*. Princeton University Press, Princeton, NJ, 2007.

R. W. Hamming. *The Art of Probability for Scientists and Engineers*. Addison-Wesley, Redwood City, CA, 1991.

B. Hansson. The independence condition on the theory of social choice. Working paper no. 2, The Mattias Fremling Society, Department of Philosophy, Lund, 1972.

J. Harsanyi. Cardinal welfare, individualistic ethics, and interpersonal comparisons of utility. *Journal of Political Economy*, 63:315, 1955.

J. Harsanyi. *Rational Behavior and Bargaining Equilibrium in Games and Social Situations*. Cambridge University Press, Cambridge, 1977.

J. Henrich et al. "Economic man" in cross-culturative perspective: Behavorial experiments in 15 small-scale societies. *Behavioral and Brain Sciences*, 28(6):795–855, 2005.

R. M. Hogarth and M. W. Reder, editors. *Rational Choice*. University Chicago Press, Chicago, 1986.

M. Hollis. *Models of Man*. Cambridge University Press, Cambridge, 1977.

S. Holmes. The secret history of self-interest. In J. J. Mansbridge, editor, *Beyond Self-Interest*, chapter 17. University of Chicago Press, Chicago, 1990.

G. C. Homans. *Social Behavior: Its Elementary Forms*. Harcourt Brace & World, New York, 1961.

G. C. Homans. *The Nature of Social Science*. Harcourt Brace & World, New York, 1967.

R. A. Howard and J. E. Matheson. Influence diagrams. In R. A. Howard and J. E. Matheson, editors, *Readings on the Principles and Applications of Decision Analysis*. Strategic Decisions Group, Menlo Park, CA, 1984.

P. Jaccard. Étude comparative de la distribution florale dans une portion des alpes et des jura. *Bulletin de la Société Vaudoise des Sciences Naturelles*, 37:547–579, 1901.

W. James. *The Will to Believe and Other Essays*. Dover, New York, 1956.

E. T. Jaynes. *Probability Theory: The Logic of Science*. Cambridge University Press, Cambridge, 2003.

R. Jeffyrey. *Probability and the Art of Judgment*. Cambridge University Press, Cambridge, 1992.

F. V. Jensen. *Bayesian Networks and Decision Graphs*. Springer Verlag, New York, 2001.

P. E. Johnson. *Social Chice: Theory and Research*. Sage Publications, Thousand Oaks, CA, 1998.

R. L. Keeney and H. Raiffa. *Decisions with Multiple Objectives*. Cambridge University Press, Cambridge, 1993. First published by John Wiley & Sons, 1976.

J. Kemeny. Fair bets and inductive probabilities. *Journal of Symbolic Logic*, 20(1): 263–273, 1955.

A. Kolmogorov. *Grundbegriffe der Warscheinlichkeitsrechnung.* Springer, Berlin, 1933.

A. Kraskov, H. Stöbauer, R. G. Andrzejak, and P. Grassberger. Hierarchical clustering based on mutual information. *ArXiv q-bio/0311039 (http://arxiv.org/abs/q-bio/0311039)*, 2003.

J. W. Krutch. *The Measure of Man.* Grosset and Dunlap, New York, 1953.

T. S. Kuhn. *The Structure of Scientific Revolutions.* University of Chicago Press, Chicago, 1962.

H. T. Kung, F. Luccio, and F. P. Preparata. On finding the maxima of a set of vectors. *Journal of the Association for Computing Machinery*, 22(4):469–476, 1975.

S. L. Lauritzen. *Graphical Models.* Springer Verlag, New York, 1996.

I. Levi. *The Enterprise of Knowledge.* MIT Press, Cambridge, MA, 1980.

I. Levi. *The Covenant of Reason.* Cambridge University Press, Cambridge, 1997.

D. K. Lewis. *Convention.* Harvard University Press, Cambridge, MA, 1969.

M. Li, J. H. Badger, X. Chen, S. Kwong, P. Kearney, and H. Zang. An information-based sequence distance and its application to whole mitochondrial genome phylogeny. *Bioinformatics*, 17(2):149–154, 2001.

R. D. Luce and H. Raiffa. *Games and Decisions.* John Wiley, New York, 1957.

T. Macaulay. Mill's essay on government: Utilitarian logic and politics. In J. Lively and J. Rees, editors, *Utilitarian Logic and Politics.* Clarendon Press, Oxford, 1978.

T. W. Malone and K. G. Crowston. Toward an interdisciplinary theory of coordination. Technical Report No. 120. Center for Coordination Science, SS WP#3294-91-MSA, MIT, 1991.

T. W. Malone and K. G. Crowston. The interdisciplinary study of coordination. In T. W. Malone, K. G. Crowston, and G. A. Herman, editors, *Organizing Business Knowledge.* MIT Press, Cambridge, MA, 2003.

H. Margolis. Dual utilities and rational choice. In J. J. Mansbridge, editor, *Beyond Self-Interest*, chapter 15. University of Chicago Press, Chicago, 1990.

J. Maynard-Smith. *Evolution and the Theory of Games.* Cambridge University Press, Cambridge, 1982.

L. Mérö. *Moral Calculations.* Springer-Verlag, New York, 1998.

J. A. H. Murray, H. Bradley, W. A. Craigie, and C. T. Onions, editors. *The Compact Oxford English Dictionary.* The Oxford University Press, Oxford, 1991.

J. Neveu. *Mathematical Foundations of the Calculus of Probability.* Holden Day, San Francisco, 1965.

M. A. Nowak, K. M. Page, and K. Sigmund. Fairness versus reason in the ultimatum game. *Science*, 289:1773–1775, 2000.

P. Ordeshook. *Game Theory and Political Theory: An Introduction.* Cambridge University Press, Cambridge, 1986.

José Ortega y Gasset. Mirabeau: An essay on the nature of statesmanship. Historical Conservation Society, Manila, 1975.

M. J. Osborne and A. Rubinstein. *A Course in Game Theory.* MIT Press, Cambridge, MA, 1994.

F. R. Palmer. *Grammar.* Harmondsworth, Middlesex, Penguin, 1971.

V. Pareto. *Manual of Political Economy.* Augustus M. Kelley, New York, 1927/1971. Translated in 1971 by Ann S. Schwier from the French edition of 1927.

J. Pearl. *Probabilistic Reasoning in Intelligent Systems*. Morgan Kaufmann, San Mateo, CA, 1988.

G. Polya. *Induction and Analogy in Mathematics*. Princeton University Press, Princeton, NJ, 1954.

K. R. Popper. *Conjectures and Refutations: The Growth of Scientific Knowledge*. Harper & Row, New York, 1963.

I. Prigogine. *Thermodynamics of Irreversible Processes*. Thomas Press, Springfield, IL, 1955.

H. Raiffa. *Decision Analysis*. Addison-Wesley, Reading, MA, 1968.

F. P. Ramsey. Truth and probability. In R. B. Braithwaite, editor, *The Foundations of Mathematics and Other Logical Essays*. The Humanities Press, New York, 1950.

E. Rasmusen. *Games and Information*. Basil Blackwell, Oxford, 1989.

J. B. Rawls. *A Theory of Justice*. Harvard University Press, Cambridge, MA, 1971.

S. Ross. *A First Course in Probability*. Prentice-Hall, Upper Saddle River, NJ, 6th ed., 2002.

S. J. Russell and P. Norvig. *Artificial intelligence: A Modern Approach*. Prentice-Hall, Upper Saddle River, NJ, 2nd ed., 2003.

T. C. Schelling. *The Strategy of Conflict*. Oxford University Press, Oxford, 1960.

A. K. Sen. Choice, ordering and morality. In S. Körner, editor, *Practical Reason*, pages 54–67. Blackwell, Oxford, 1979a.

A. K. Sen. *Collective Choice and Social Welfare*. North-Holland, Amsterdam, 1979b.

A. K. Sen. Rational fools: A critique of the behavorial foundations of economic theory. In J. J. Mansbridge, editor, *Beyond Self-Interest*, chapter 2. University of Chicago Press, Chicago, 1990.

G. Shafer. *A Mathematical Theory of Evidence*. Princeton University Press, Princeton, NJ, 1976.

C. Shannon and W. Weaver. *The Mathematical Study of Communication*. University of Illinois Press, Urbana, IL, 1949.

Y. Shoham and K. Leyton-Brown. *Multiagent Systems*. Cambridge University Press, Cambridge, 2009.

M. Shubik. *Game Theory in the Social Sciences*. MIT Press, Cambridge, MA, 1982.

M. Shubik. Game theory and operations research: Some musings 50 years later. Yale School of Management Working Paper No. ES-14, 2001.

K. Sigmund, E. Fehr, and M. A. Nowak. The economics of fair play. *Scientific American*, 286:83–87, 2002.

H. A. Simon. A behavioral model of rational choice. *Quartly Journal of Economics*, 59: 99–118, 1955.

H. A. Simon. Rationality in psychology and economics. In R. M. Hogarth and M. W. Reder, editors, *Rational Choice*. University Chicago Press, Chicago, 1986.

H. A. Simon. *The Sciences of the Artificial*. MIT Press, Cambridge, MA, 3rd ed., 1996.

M. Slatkin. Altruism in theory [review of Scott Boorman and Paul Levitt, *The Genetics of Altruism*]. *Science*, 210(4470):633–634, 1980.

M. Slote. *Beyond Optimizing*. Harvard University Press, Cambridge, MA, 1989.

E. Sober and D. S. Wilson. *Unto Others: The Evolution and Psychology of Unselfish Behavior*. Harvard University Press, Cambridge, MA, 1998.

I. Steedman and U. Krause. Goethe's *Faust*, Arrow's Possibility Theorem and the individual decision maker. In J. Elster, editor, *The Multiple Self*, chapter 8, pages 197–231. Cambridge University Press, Cambridge, 1985.

W. C. Stirling. *Satisficing Games and Decision Making: With Applications to Engineering and Computer Science*. Cambridge University Press, Cambridge, 2003.

M. Tribus. *Rational Descriptions, Decisions and Designs*. Pergamon Press, New York, 1969.

A. Tversky and D. Kahenman. Rational choice and the framing of decisions. In R. M. Hogarth and M. W. Reder, editors, *Rational Choice*, pages 67–94. University of Chicago Press, Chicago, 1986.

J. von Neumann and O. Morgenstern. *The Theory of Games and Economic Behavior*. Princeton University Press, Princeton, NJ, 1944. 2nd ed., 1947.

J. von Plato. *Creating Modern Probability*. Cambridge University Press, Cambridge, 1994.

M. Weber. *Economy and Society*. University of California Press, Berkeley, 1968. Edited by G. Roth and C. Wittich.

J. W. Weibull. *Evolutionary Game Theory*. MIT Press, Cambridge, MA, 1995.

G. Wolf and M. Shubik. Concepts, theories and techniques: Solution concepts and psychological motivations in prisoner's dilemma games. *Decision Sciences*, 5:153–163, 1974.

L. A. Zadeh. What is optimal? *IRE Transactions on Information Theory*, 4(1):3, 1958.

L. A. Zadeh. Fuzzy sets as a basis for a theory of possibility. *Fuzzy Sets and Systems*, 1:3–28, 1978.

Name Index

Subject Index

acceptability, 143
action interdependence, 98, 100
action profile, 2, *33*
action space, 33
acyclicity, 53–54, 58, 68, 160
affine transformation, 51
aggregation
  cardinal, 49
  endogenous, 58, 59
  function, 55, 59
    epistemic, 48
    praxeic, 49
  ordinal, 47
  theorem, 59
agnosticism, 63
agree, 51
altruism, viii, 6, 20, 155
  categorical, 38, 204
  conditional, 38, 176, 204
ambivalence, 107
analysis, ix, 5
Arrow's impossibility theorem, 4, 27, 36,
  101, 139
artificial intelligence, x, 71, 97, 103, 125, 210
aspiration level, 157, 159
associativity equation, 60
asymmetry, 89
atoms, 156

Battle of the Sexes, 21, 174
Bayes rule, 63, 69
binary preference relation, 33
  antisymmetry, 33
  asymmetry, 33
  complete ordering, 33

indifference, 33
partial ordering, 33
reflexivity, 33
strict ordering, 33
strict partial ordering, 33
symmetry, 33
total ordering, 33
transitivity, 33
boldness, 148, 158, 163
Boolean algebra, 145, 146

categorical preference, 19, 32, 35
categorical preference ordering, viii, 3, *3*, 5, 7,
  10, 13, 14, 17, 20
categorical subjugation, 66
categorical utility, *52*
certainty-equivalence hypothesis, 126
chain rule
  hardness, 114
  praxeic, 61, 63
  probability, 50
chance node, 127
coherence, 64
  epistemological, 64
  praxeological, 65, 66, 143
  satisficing, 164, 166
competition, 1, 4, 5, 22, 100, 101, 103, 112
compromise, viii, x, 4, 5, 20, 78, 79, 84, 91,
  94, 100, 112, 125, 154, 156, 158
  satisficing, 169
compromise set, *79*, 92, 94, 169
  satisficing, 167, 178, 189
  satisficing social, 168
computer science, ix–x, 1, 5, 26
concordance, ix, 19, 39, 40